Translational Approaches to Autism Spectrum Disorder

Maria de los Angeles Robinson-Agramonte
Editor

Translational Approaches to Autism Spectrum Disorder

Springer

Editor
Maria de los Angeles Robinson-Agramonte
Neuroimmunology Department
International Center for Neurological
Restoration
Havana, Cuba

ISBN 978-3-319-16320-8 ISBN 978-3-319-16321-5 (eBook)
DOI 10.1007/978-3-319-16321-5

Library of Congress Control Number: 2015939345

Springer Cham Heidelberg New York Dordrecht London

Printed on acid-free paper

Springer is part of Springer Science+Business Media (www.springer.com)

I do not know where we go,
I alone want to go with you

To my father's memory,

To my mother

To my son

Preface

The publication Translational Approaches to Autism Spectrum Disorder (ASD) combines an important synthesis of clinical and experimental views from the molecular to the behavioral, of this disease that constitutes a real challenge to neurology, psychiatry, psychology, and basic neuroscience. During a meeting in Cancun, Mexico (April 21–25, 2013), the main hypotheses of ASD were intensively discussed by experts from several countries covering all the levels of neuroscience (molecular to behavioral) and branches (from genetics, immunology, clinical neurophysiology, and pharmacology to epidemiology). This book is an account of the spirit of the discussion realized over those days.

The promise that translational medicine will improve the lives of children suffering from ASD is greater than ever. Knowledge of the development and behavior of ASD is increasing at a great rate, from the molecular to sociological level. These exciting scientific advances in the description of the ASDs, although incomplete so far, provide hope to effectively address new treatments for the people suffering from the illness.

Unfortunately, ASD has many variants and the presence of a real specific marker is still unknown, preventing the discovery of successful personalized treatment. This book has the potential to help researchers and clinicians to unite to uncover the clues of ASD. To fully understand this idea, it is recommended that the chapters are read in three interconnected parts: Part 1, composed of the first three chapters; Part 2, Chaps. 4–10, and Part 3, corresponding to Chap. 11.

Part 1 synthesizes the history, prevalence, and possible causes of ASD (Chap. 1); the prevalence of ASD since 2000 and the discussion of methodological factors impacting the estimation of prevalence (Chap. 2); and finally in Chap. 3, the presentation of several genetic syndromes of ASD as a multifactor and complex condition with a marked genetic influence, evidenced by a high heritability (80–90 %). Both common and rare genetic variants have an influence on the etiology and development of ASD.

Part 2 contains several approaches. At first, reviews of EEG findings with an emphasis on quantitative measurements, epilepsy, sleep disorders (Chap. 4), and dysfunction of auditory, visual, and somatosensory systems that underlie ASD (Chap. 5) are discussed. Then in Chap. 6, an overview of the morphological and functional brain changes in ASD using MRI and fMRI observations in ASD is provided. Afterwards, evidence derived from animal experiments and clinical data, supporting the role of neuroplasticity on autism physiopathology are discussed (Chap. 7). This part ends with the presentation of immunological data: first, the evidence that certain immune stimuli result in reduced sociability and increased repetitive behavior in animal models (Chap. 8); later, in Chap. 9, how immunological derangements, including cellular immune dysregulations, chronic inflammatory states, and neuroimmune alterations occur in the periphery and in the brain of those suffering from ASD. The same is true for studying antibody responses in the brain and periphery of ASD patients (Chap. 10).

Part 3 is covered entirely in Chap. 11. This section starts with the consideration that although common symptoms are well described in ASD, there is an inadequate understanding of the mechanisms at molecular and cellular levels. Thereafter, the current therapies are described, including physiological interventions, occupational and physical therapy, speech and language therapy, medications and sensory integration, and vision therapy. Other therapies discussed for the future are: intranasal oxytocin, bioactive peptides, and vitamins. It is finally concluded that a definitive pharmacologic treatment for the core symptoms of autism does not exist.

The outstanding value of this book can be attributed to its expert and authoritative contributors. Students of the field of ASD will be indebted to these dedicated authors for their hard work, knowledge, thoughtfulness, and good judgment. My sincere appreciation

MD Nibaldo Hernández Mesa
Dr. Sc. Emeritus Professor

Acknowledgments

The friends are angels that help us
To stand up when our wings forget flying
I want to offer my biggest gratefulness to the authors contributing to this book.
Thank you very much

Contents

Contributors

Nicola Antonucci Biomedical Centre for Autism Research and Treatment, Bari, Italy

Margarita Minou Báez Martin International Center for Neurological Restoration (CIREN), Clinical Neurophysiology Service, Havana, Cuba

Karla Batista Brain Images Processing Group, International Center for Neurological Restoration, Havana, Cuba

Jorge Bergado-Rosado Department of Experimental Neurophisiology, International Center for Neurological Restorations, Havana, Cuba

Ivette Cabrera Abreu International Center for Neurological Restoration (CIREN), Clinical Neurophysiology Service, Havana, Cuba

Maylen Carballo Brain Images Processing Group, International Center for Neurological Restoration, Havana, Cuba

Laura de Magistris Gastroenterology, Department Magrassi-Lanzara, Second University of Naples, SUN, Naples, Italy

Amaicha Mara Depino Institute of Physiology, Molecular Biology and Neuroscience, CONICET—Department of Physiology, Molecular and Cellular Biology, Faculty of Exact and Natural Sciences, University of Buenos Aires, Buenos Aires, Argentina

Eric Fombonne Department of Psychiatry, Autism Research, Institute for Development & Disability, Oregon Health & Science University, Portland, OR, USA

Maria Elena Gonzàlez Fraguela Neurobiochemistry Lab, Department of Immunochemistry, International Center for Neurological Restoration, Havana, Cuba

Nadia Kazlauskas Institute of Physiology, Molecular Biology and Neuroscience, CONICET—Department of Physiology, Molecular and Cellular Biology, Faculty of Exact and Natural Sciences, University of Buenos Aires, Buenos Aires, Argentina

Gerry Leisman The National Institute for Brain and Rehabilitation Sciences, Nazareth, Israel

O.R.T.-Bruade College of Engineering, Karmiel, Israel

Calixto Machado Neurophysiology Department, Institute of Neurology and Neurosurgery, Havana, Cuba

Roberta Maresca Department of Mental and Physical Health and Preventive Medicine, Second University of Naples, SUN, Naples, Italy

Luiza Monteavaro Mariath Genetics Department, Universidade Federal do Rio Grande do Sul, Porto Alegre, Brazil

Lilia María Morales Chacón International Center for Neurological Restoration (CIREN), Clinical Neurophysiology Service, Havana, Cuba

Elena Noris-García Psychoneuroimmunology Department, Nephrology Institute, Medical Havana University, Havana, Cuba

Alison Presmanes Hill Research Department of Pediatrics, Oregon Health & Science University, Portland, OR, USA

Maria de los Angeles Robinson-Agramonte Neuroimmunology Department, International Center for Neurological Restoration, Havana, Cuba

Mercedes Adalys Rodríguez-Ravelo Endocrinology Laboratory Ameijeiras Brothers Hospital, Medical Havana University, Havana, Cuba

Rafael Rodríguez-Rojas Brain Images Processing Group, International Center for Neurological Restoration, Havana, Cuba

Tatiana Roman Genetics Department, Universidade Federal do Rio Grande do Sul, Porto Alegre, Brazil

Jaqueline Bohrer Schuch Genetics Department, Universidade Federal do Rio Grande do Sul, Porto Alegre, Brazil

Lavinia Schuler-Faccini Genetics Department, Universidade Federal do Rio Grande do Sul, Porto Alegre, Brazil

Gustavo Sierra BioCubaFarma: Group of Biotechnological and Pharmaceutical Industries of Cuba, Havana, Cuba

Dario Siniscalco Department of Experimental Medicine—Division of Pharmacology, Second University of Naples, Caserta, Italy

Yamila Adams Villalón Institute of Hematology and Immunology, La Habana, Cuba

Katharine Zuckerman Division of General Pediatrics, Oregon Health & Science University, Portland, OR, USA

Chapter 1
Autism: What Is It?

Roberta Maresca and Laura de Magistris

Abstract An overview of autism is provided. History, prevalence and causes are described. The autism spectrum disorders (ASD) are developmental disorders defined by significantly abnormal social interaction, impaired communication and language abilities and a narrow pattern of interests. According to the Diagnostic and Statistical Manual of Mental Disorders (IV edition-text revision), autistic disorder (AD) is the most severe form of ASD. Asperger syndrome, Rett syndrome, childhood disintegrative disorder and pervasive developmental disorder not otherwise specified are considered milder forms of ASD. The American Psychiatric Association has recently released the fifth edition of the Diagnostic and Statistical Manual of Mental Disorders (DSM-V). The diagnostic criteria for ASD have been modified based on the research literature and clinical experience in the 19 years since the DSM-IV was published in 1994. In particular, the diagnosis will be called Autism Spectrum Disorder and there will no longer be sub-diagnoses.

Keywords History · Diagnosis · Prevalence · Causes · Social impact · Autism

1.1 Introduction

The prevalence of autism spectrum disorders (ASD) is increasing and it is alarming. In fact, 1 in 88 newborns is currently affected by ASD in the USA, with a specific male dominance (5:1 ratio of males to females). The presence of a child with ASD in the family causes an overall impact on parents and siblings manifested in a significant increase in stress.

L. de Magistris (✉)
Gastroenterology, Department Magrassi-Lanzara, Second University of Naples,
SUN, Naples, Italy
e-mail: laura.demagistris@unina2.it

R. Maresca
Department of Mental and Physical Health and Preventive Medicine,
Second University of Naples, SUN, Naples, Italy

M. d. l. A. Robinson-Agramonte (ed.), *Translational Approaches to Autism Spectrum Disorder*, DOI 10.1007/978-3-319-16321-5_1

Only about 10 % of patients with a diagnosis of ASD have a defined aetiology (so-called syndromic autism, secondary to fragile X syndrome, neurofibromatosis, exposure to thalidomide), while 90 % of ASD cases are considered idiopathic (i.e. without a definite aetiological agent). In these cases, genetic factors are known to be relevant. However, genetics only cannot explain the rapid increase of the prevalence observed in recent years. For this reason, most authors consider that the aetiology of ASD lies in an interaction between genetic and environmental factors. The genes that have so far been identified seem to have an important role to play in brain development, particularly in synaptic development. The results of the screening of the human genome have shown the presence of different susceptibility loci on different chromosomes, in particular the 2, 16 and 17. The data obtained suggest the interaction of at least ten genes in the aetiology of autism. In practice, the strongest evidence is that there is no 'one gene' of autism, but quite a number of genes that contribute to a vulnerability to the onset of the disorder. There are many challenging environmental factors that appear to play a role in autism. It is unclear how these environmental factors (including vaccines) may interact with the individual genetic risk for developing autism. Research has not yet identified a direct causal link between any environmental factor and autism.

1.2 History and Diagnosis

The term autism was coined in 1908 by the Swiss psychiatrist Bleuler to refer to a symptom of schizophrenia, pointing to it with the extreme narrowing of relations with the outside world that excludes anything except your own self: it derives from the Greek αυτός, or "if the same".

The first scientific description of autism—a psychopathological syndrome in itself—is due to the German-American physician Leo Kanner in 1943, in the article "Autistic Disturbances of Affective Contact" published in *The Nervous Child*. He described eleven children prone to isolation, "self-sufficient", "happy if left alone", "as in a shell", not very reactive in the relational field (Paucheri and Pfanner 1999). Some appeared functionally silent or had echolalic speech, while others showed a characteristic inversion pronoun ("you" to refer to themselves and "I" to refer to another). Many had an obsessive fear of some change happening in the surrounding environment, while some had very specific skills, developed and isolated, but close to a general delay. Observing these traits and behaviours as typical but hitherto unknown, Kanner described this syndrome as "Early Infantile Autism". Although experts in autism have been the first to recognize, under the same name, a clinical condition characterized by precise impaired social interaction, impaired communication and the constant presence of repetitive, stereotyped and restricted interests, and have changed the diagnostic and therapeutic approach over time, there are still large areas of uncertainty, doubt and problem.

In fact, numerous efforts to understand the causes of this syndrome and various assumptions have been made; the multifactorial origin, which needs the interaction of genetic and environmental factors, seems however to be the most reliable.

Autism is a behavioural syndrome caused by a developmental disorder that is biologically determined, with onset in the first 3 years of life. The areas relating to social interaction, the capacity to communicate ideas and feelings and the ability to establish relationships with others are mainly affected; therefore, it appears as a "permanent" disability that accompanies the subject through his lifetime, although the characteristics of the social deficit assume variable expressivity in time.

According to the definition of the American Psychiatric Association, significantly abnormal or deficient social interaction, impaired communication and language abilities and a considerably narrow pattern of activities and interests are the mainstays of the ASD diagnosis.

ASDs, in the DSM-V, are defined within two categories: "persistent impairment in reciprocal social communication and social interactions" and "restricted and repetitive patterns of behaviour," both present since early childhood. In the DSM-V diagnosis must meet the following criteria A, B, C and D:

a. Persistent deficits in social communication and social interaction across contexts, not accounted for by general developmental delays, and manifested by all three of the following:

    1. Deficits in social–emotional reciprocity; ranging from abnormal social approach and failure of normal back-and-forth conversation, through reduced sharing of interests, emotions, and affect and response, to total lack of initiation of social interaction.
    2. Deficits in nonverbal communicative behaviours used for social interaction; ranging from poorly integrated verbal and nonverbal communication, through abnormalities in eye contact and body language, or deficits in understanding and use of nonverbal communication, to total lack of facial expression or gestures.
    3. Deficits in developing and maintaining relationships appropriate to the level of development (beyond those with caregivers); ranging from difficulties adjusting behaviour to suit different social contexts, through difficulties in sharing imaginative play and in making friends, to an apparent absence of interest in people.

b. Restricted, repetitive patterns of behaviour, interests, or activities as manifested by at least two of the following:

    1. Stereotyped or repetitive speech, motor movements, or use of objects; (e.g. simple motor stereotypes, echolalia, repetitive use of objects or idiosyncratic phrases).
    2. Excessive adherence to routines, ritualized patterns of verbal or nonverbal behaviour, or excessive resistance to change; (e.g. motoric rituals, insistence on same route or food, repetitive questioning or extreme distress at small changes).

3. Highly restricted, fixated interests that are abnormal in intensity or focus; (e.g. strong attachment to or preoccupation with unusual objects, excessively circumscribed or perseverative interests).
4. Hyper- or hypo-reactivity to sensory input or unusual interest in sensory aspects of the environment; (e.g. apparent indifference to pain/heat/cold, adverse response to specific sounds or textures, excessive smelling or touching of objects, fascination with lights or spinning objects).

c. Symptoms must be present in early childhood (but may not become fully manifest until social demands exceed limited capacities)
d. Symptoms together limit and impair everyday functioning.

The diagnosis also requires the specification of the presence or absence of related intellectual disability, alterations of language as well as medical conditions or associated genetics.

1.3 Prevalence

The prevalence of ASD is increasing and the current prevalence rate is alarming. In fact one in 88 newborns is currently affected by ASD in the USA, with a specific male dominance (5:1 ratio) between males and females (Baio 2012). However, a recent study performed in the UK suggests lower figures (annual incidence rate of about 1.2/1000 boys and 0.2/1000 girls). Only about 10% of patients with a diagnosis of ASD have a defined aetiology (the so-called syndromic autism, previously defined; Engel and Daniels 2011; Stagno and Whitley 1985), while 90% of ASD cases are considered idiopathic (i.e. without a definite aetiological agent) (Gentile et al. 2013). In these cases, genetic factors are known to be relevant. However, genetics cannot explain the rapid increase of the prevalence observed in recent years. For this reason, most authors consider that the aetiology of ASD lies in an interaction between genetic and environmental factors (Garbett et al. 2012; Shi et al. 2003). The absence of a definite causative agent makes the set-up of preventive measures for ASD impossible.

1.4 Causes of Autism

In the 1940s, Kanner described how the parents of his patients with autism were able to identify a characteristic common to all of them: that of being "cold and intellectual". Afterwards, in his theory of psychoanalysis, Bettelheim affirmed that "the refusal on the part of the parents is a key element in the genesis of each observed case of autism" (Bettelheim 1967). In practice, he opened the way to what has been, for decades, the cornerstone of the interpretation of the onset of autistic disorder, namely the theory of the "refrigerator mother". Emphasizing the absence of an organic background and the particular type of parenting, autism was interpreted as a defense against anxiety, arising from a failure of the first object relations. This model, however, was not broadly accepted everywhere, and in the Anglo-Saxon countries psychoanalytic hypotheses never found wide scientific credit. Bernard Rimland, an American psychologist father of an autistic child, revolutionizes the approach to this disease, arguing that the problem was not to be found in the "parents" of the child, but rather in an organic cause: it was 1960. Since then different theories have started being processed; among them the socio-affective theory of mind and the executive function theory; the input to the search of a matrix within the neurobiological disorder began to materialize.

The neurobiological research in the area of ASD starts with the "Theory of Socio-Affective" (Hobson 1993). This research area was aimed at defining the characteristics of mental functioning autistics, from which the behaviours that characterize the clinical picture should be derived. According to this theory, the autistic child has an innate inability, biologically determined (the existence of brain structures responsible for processing of social stimuli are assumed), to interact with other people emotionally. The absence of this empathy is not referential (Hobson and Lee 1989) or primary inter-subjective (Trevarthen and Aitken 2001): it is due to a sort of cascade reaction leading to an inability to learn and/or recognize the mental states of others, to impaired symbolization, to the deficit of language and of social interaction.

Other assumptions in this research direction limit their attention to the so-called central coherence and executive function. Central coherence refers to the ability to synthesize into a coherent whole the various experiences that affect senses in a fragmented way (Frith and Happé 1994). A weakness of this aspect would therefore be to remain at parceled experiential data, not allowed to grasp the ultimate meaning of each stimulus, just as in autism. The concept of executive function is also very similar to the previous (Bennetto et al. 1996). Indeed, there are skills that are crucial in the organization and planning of behaviours for the resolution of problems. This refers to the ability to activate an area of mental work, mentally formulate a plan of action, not to be rigidly anchored in the perceptual data, to inhibit impulsive responses and to be attentive to the feedback information in order to move flexible attention to the various aspects of the context. It is clear that a deficiency of these aspects leads precisely to events like impulsiveness, inability to inhibit appropriate responses, hyper-selectivity, remaining anchored to detail and perseverance: all these aspects are peculiarly close to autism.

The theory of mind (Baron-Cohen et al. 2000) was formulated in the eighties; it assumes that the child begins to acquire the ability to reflect on emotions, desires and beliefs about self and others at 4 years of age. If this does not happen, the child does not gain the ability to predict the behaviour of others, and will not be able to accomplish the level of "meta-representations" (does not know how to think that other people think and what they think). This translates into hyper-selectivity, repetitiveness, rigidity and perseveration and will possibly affect the area of social interaction and communication. The recent discovery of the mirror system (Cattaneo et al. 2007) suggests a "physiological" explanation to this theory. The mirror system is a neural system present in humans and non-human primates useful for understanding imitation and the intentions of others. The organization of intentional motor chains in children with typical development and with autism, by means of electromyography (EMG) was investigated; it was demonstrated that the intentional chains are altered in autistic children (Cattaneo et al. 2007). The high-functioning autistic children understand the intentions of others cognitively, but lack a mechanism allowing direct and immediate experiential understanding of the action. High functioning autism is, in fact, characterized by preserved intelligence and general cognitive functioning accompanied by deficits in social interaction and communication, as well as repetitive and restrictive behaviour.

As a matter of fact, the causes of ASD are still unknown. Other hypotheses include genetics, obstetric complications, infection and toxic exposures (Larsson et al. 2005; Lawler et al. 2004; Constantino et al. 2010). None of these, however, has been established as a definite aetiological agent.

Familial factors influence the risk for autism spectrum disorder. The rate of ASD in children born into families that already have an affected child is as high as 18.7 %, and the risk is twice as high in children born into families with two or more affected children (Ozonoff et al. 2011). Girls born into a family that has a sick child with an ASD will have 2.8 more times the risk of having such a disorder (Ozonoff et al. 2011). Twin studies have demonstrated a moderate degree of genetic heritability for ASD (Abrahams and Geschwind 2008, 2010; Hallmayer et al. 2011), with environment making a substantial contribution to the development of these conditions in the study subjects (Hallmayer et al. 2011). It was noticed that identical twins have a higher probability (concordance rate higher than 60 %) of both being autistic compared to non-identical twins, and that heterozygous parents of an affected child have a recurrence rate of having a second child with an ASD that varies between 2 and 8 %, in contrast to the prevalence in the general population that is around 3 %. Family members of patients who have similar though faded behavioural characteristics were also identified. Finally, the presence of a secondary or syndromic autism associated with genetically determined diseases such as Tuberous Sclerosis, Fragile X Syndrome, and Neurofibromatosis is significant, thus confirming the influence of the genetic component in the onset of the disorder (Muhle et al. 2004).

The results of the screening of the human genome have shown the presence of different susceptibility loci on different chromosomes, in particular the 2, 7, 16 and 17. Obtained data suggest the interaction of at least ten genes in the aetiology of autism. In practice, the evidence that there is no "one-gene" autism is strong; quite a number of genes, however, contribute to a vulnerability to the onset of the disorder. Karyotype analysis also showed deletions and/or duplications in the long arm of chromosome 15 and 22 (Borgatti et al. 2001).

Many individuals with autism and related conditions experienced untoward events in their prenatal and neonatal periods and during delivery (Brasic and Holland 2006, 2007; Brasic et al. 2003; Glasson et al. 2004). It is unclear whether the obstetric complications caused autistic disorder or whether autism and obstetric complications resulted from environmental or other problems. In particular, the consumption of valproate or thalidomide during pregnancy leads to brain abnormalities that include the absence of the motor nuclei of cranial nerves and a reduction of the "brainstem" (Narita et al. 2002).

It can therefore be said that no significant association between pathogenic noxae, maternal medical conditions such as pregnancy, childbirth or other problems related to environmental factors and autism has been demonstrated.

Some studies (Gillberg et al. 1990, 1992; Gillberg and Coleman 1992) have shown a higher incidence of perinatal diseases in populations of individuals with autism compared to control groups; this reinforces the hypothesis that subjects with genetically determined disorders have a low power to be born, that predisposes them to a perinatal pre-suffering.

Both paraoxonase (directed against the diazoxon, one of the major components of organophosphates) and arylesterase (antioxidant role against LDL) activities, with particular relevance to the second, are reduced in the course of autism (D'Amelio et al. 2005). The paraoxonase is an enzyme encoded by the PON1 gene, synthesized by the liver and transported in the plasma bound to HDL. The interesting fact is that the measurement of arylesterase activity, together with the characterization of polymorphism in the PON1, allows distinguishing autism from controls and controls by parents, the last appearing as a "healthy carrier" of the pathology. Therefore, the implementation of this simple biochemical/genetic test could provide a distinction between affected and unaffected subjects. It is very interesting, however, that a similar enzyme activity reduction has been detected in humans and in animal models, only in the course of viral infection accompanied by an intense immune reaction (Lenten et al. 2002; Kilic et al. 2005; Parra et al. 2007). These results are in keeping with another important pathogenetic hypothesis of autism: the connection with a deregulated immune system (more data in Chap. 9). The various gene associations, initially interpreted as linked to neurodevelopment, could instead be mediated in whole or at least in part by the immune system. This applies, for example, to the MET gene (7q31), which encodes a receptor tyrosine kinase, involved in the growth and maturation of neocortex and cerebellum, immune function and repair in the gastrointestinal tract, and which levels appear reduced in patients with autistic disorder (Campbell et al. 2006, 2007). Interestingly, a correlation between the presence of macrocephaly and a previous history of allergic disorders/immunity was shown (Sacco et al. 2007).

Attention was placed on the relationship between vaccination and the onset of some autistic behaviours. At present, however, there are no data to suggest that any vaccine increases the risk of developing autism or any other disorders of behaviour, nor it is established whether the risk of occurrence of this framework is related to the vaccine itself and its action on the immune system or to the salt of ethylmercury used as a preservative in the preparation of vaccines (Bernard et al. 2002).

Meta-analyses of epidemiologic studies have shown that autism risk in offspring increases with the advancing age of either parent. Sandin et al. reported that the adjusted relative risk for autism was 1.52 in the offspring of mothers aged 35 years or older compared with mothers aged 25–29 years (Sandin et al. 2012). Hultman et al. found that offspring of men aged 50 years or older were 2.2 times more likely to have autism than offspring of men aged 29 years or younger (Hultman et al. 2011).

A final consideration has to be made about the circadian rhythms of people with autism, which often are altered when compared to those of unaffected subjects. Among sleep disorders, the noise of falling, as well as nocturnal awakenings, restless and disturbed sleep, are particularly common anamnestic data to be collected. This appears to be linked to a deficiency of the hormone melatonin, which production depends on the suprachiasmatic nucleus, and that is responsible for alternating sleep/wakefulness. It was shown that the ASMT gene, encoding the last enzyme in the synthetic pathway of melatonin, is deleted in some autistic individuals (Taft and Cohen 1971; Wakefield et al. 1998).

Consistent data on associated gastrointestinal symptoms and increased intestinal permeability have been reported in the literature. An inflammatory bowel disease, named “autistic enterocolitis”, characterized by a chronic ileo-colitis with ileal lymphoid nodular hyperplasia associated with a high degree of systemic framework complicated by signs of immunodeficiency and autoimmunity, was reported for the first time by Wakefield et al. more than 10 years ago (Bernard et al. 2002; Wakefield et al. 1998). The most common associated symptoms, variously depending on the subject, are constipation, diarrhoea with steatorrhoea and macroscopic signs of malabsorption, alternating bowel with constipation/diarrhoea, intestinal dysbiosis, abdominal distension, digestive enzyme insufficiency, gastro esophageal reflux, vomiting food, abdominal pain, mega rectum, pathological thinness with a reduction in lean body mass, poor weight gain (Buie et al. 2010). The presence of such GI involvement could explain certain peculiar clinical conditions (antalgic positions, abdominal peculiar position, evacuative atypical with increased abdominal pressure, disimmunity with particular sensitivity to gastrointestinal infections) or eating habits (anorexia/food selection on the basis of consistency, fluid intake, autoredox dairy/carbs), that in the past were described as common stereotypes related to the pathology on a neuropsychiatric basis. In patients with autistic enterocolitis, high activation of IL-2, IL-8 and M-CSF is present as compared to healthy controls (Jyonouchi et al. 2001; Gupta et al. 1998). It has been demonstrated that the signal of cytokines gene activation occurs only in the GALT (gastrointestinal associated lymphoid tissue) and therefore the gut is the main source of onset of the immune cascade. Recent studies have confirmed that the intestinal pathology can be a source of toxic or immunological damage to the brain during development (Coury et al. 2012). This possibility is reinforced when a leaky barrier condition is present, as demonstrated by altered intestinal permeability (De Magistris et al. 2010, 2013).

The abnormal activation of innate immune response (not antigen-specific) was shown in some autistic patients. This may represent an inflammatory reaction secondary to the cerebral innate immune system following stimulation that occurred elsewhere. Vargas and colleagues identified this locus in leptomeninge and in the choroid plexus; the detection of differences, both qualitatively and quantitatively, were obtained examining the cytokine profile of extracts of brain tissue samples and cerebrospinal fluid (Coury et al. 2012. The reasons for these differences remain unresolved; the main hypothesis is that the gut is the main source of activation of the immune system genes which effect could include remote activation mechanisms of the innate immune system in the brain, contributing to the onset of psychiatric symptoms characteristic of ASD.

The results of the studies reported above, taken together, make it possible to envisage autism as the result of a disconnection of distant brain areas; these could arise from genetic mutations and chromosomal rearrangements, which directly interfere with synaptogenesis, or result from an early and abnormal activation of the immune system starting from either the nervous system or the digestive tract, or both. The reasons for widespread development of ASD remain, unfortunately, an open chapter giving strong reasons to continue investigating.

1.5 Social Impact

For an individual, the impact of having autism is that they often experience failure in school, social and work situations. This leads to lack of confidence and low self-esteem. For many, it leads to high anxiety, depression and mental health difficulties. Many people with autism are also very vulnerable to abuse because of their social deficit. Being bullied and taken advantage of by so-called "friends" is not unusual. Caring for an autistic child or young adult can be a tremendous emotional, financial and physical strain with a dramatic influence in society.

References

Abrahams BS, Geschwind DH (2008) Advances in autism genetics: on the threshold of a new neurobiology. Nat Rev Genet 9(5):341–355

Abrahams BS, Geschwind DH (2010) Connecting genes to brain in the autism spectrum disorders. Arch Neurol 67(4):395–399

Baio J (2012) Prevalence of autism spectrum disorders. Autism and developmental disabilities monitoring network, 14 sites, United States, 2008. MMWR Surveill Summ 61:1–19

Baron-Cohen S, Ring HA, Bullmore ET, Wheelwright S, Ashwin C, Williams SC (2000) The amygdala theory of autism. Neurosci Biobehav Rev 24(3):355–364

Bennetto L, Pennington BF, Rogers SJ (1996) Intact and impaired memory functions in autism. Child Dev 67(4):1816–1835

Bernard S, Enayati A, Roger H, Binstock T, Redwood L (2002) The role of mercury in the pathogenesis of autism. Mol Psychiatry 2(Suppl 7):S42–43

Bettelheim B (1967) The empty fortress. Macmillan Company, New York (Italian: La Fortezza Vuota, Feltrinelli, Milano 1976)

Borgatti R, Piccinelli P, Passoni D, Dalprà L, Miozzo M, Micheli R, Gagliardi C, Balottin U (2001) Relationship between clinical and genetic features in "inverted duplicated chromosome 15" patients. Pediatr Neurol 24(2):111–116

Brasic JR, Holland JA (2006) Reliable classification of case-control studies of autistic disorder and obstetric complications. J Dev Phys Disabil 18:355–381

Brasic JR, Holland JA (2007) A qualitative and quantitative review of obstetric complications and autistic disorder. J Dev Phys Disabil 19:337–364

Brasic JR, Holland JA, Alexander M (2003) The increased likelihood of obstetric complications in autistic disorder. South Med J 96(Suppl 10):S34

Buie T, Campbell DB, Fuchs GJ III, Furuta GT, Levy J et al (2010) Evaluation, diagnosis, and treatment of gastrointestinal disorders: a consensus report. Pediatrics 125:S1–S18

Campbell DB, Sutcliffe JS, Ebert PJ, Militerni R, Bravaccio C, Trillo S, Elia M, Schneider C, Melmed R, Sacco R, Persico AM, Levitt P (2006) A genetic variant that disrupts MET transcription is associated with autism. Proc Natl Acad Sci U S A 103(45):16834–16839

Campbell DB, D'Oronzio R, Garbett K, Ebert PJ, Mirnics K, Levitt P, Persico AM (2007) Disruption of cerebral cortex MET signaling in autism spectrum disorder. Ann Neurol 62(3):243–250

Cattaneo L, Fabbri-Destro M, Boria S, Pieraccini C, Monti A, Cossu G, Rizzolatti G (2007) Impairment of actions chains in autism and its possible role in intention understanding. Proc Natl Acad Sci U S A 104(45):17825–17830

Constantino JN, Zhang Y, Frazier T, Abbacchi AM, Law P (2010) Sibling recurrence and the genetic epidemiology of autism. Am J Psychiatry 167(11):1349–1356

Coury DL, Ashwood P, Fasano A et al (2012) Gastrointestinal conditions in children with autism spectrum disorder: developing a research agenda. Pediatrics 130:S160–S168

D'Amelio M, Ricci I, Sacco R, Liu X, D'Agruma L, Muscarella LA, Guarnieri V, Militerni R, Bravaccio C, Elia M, Schneider C, Melmed R, Trillo S, Pascucci T, Puglisi-Allegra S, Reichelt KL, Macciardi F, Holden JJ, Persico AM (2005) Paraoxonase gene variants are associated with autism in North America, but not in Italy: possible regional specificity in gene-environment interactions. Mol Psychiatry 10(11):1006–1016

De Magistris L, Familiari V, Pascotto A, Sapone A, Frolli A, Iardino P, Carteni M., De Rosa M, Francavilla R, Riegler G, Militerni R, Bravaccio C (2010) Alterations of the intestinal barrier in autistic patients and in their first degree relatives. J Pediatr Gastroenterol Nutr 51:418–424

De Magistris L, Picardi A, Siniscalco D, Riccio MP, Sapone A, Cariello R, Abbadessa S, Medici N, Lammers KM, Schiraldi C, Iardino P, Marotta R, Tolone C, Fasano A, Pascotto A, Bravaccio C (2013) Antibodies against food antigens in patients with autistic spectrum disorders (ASDs). Biomed Res Int 2013:729349

Engel SM, Daniels JL (2011) On the complex relationship between genes and environment in the etiology of autism. Epidemiology 22:486–488

Frith U, Happé F (1994) Language and communication in autistic disorders. Philos Trans R Soc Lond B Biol Sci 346(1315):97–104

Garbett KA, Hsiao EY, Kalman S, Patterson PH, Mirnics K (2012) Effects of maternal immune activation on gene expression patterns in the fetal brain. Transl Psychiatry 2:e98

Gentile I, Zappulo E, Militerni R, Pascotto A, Borgia G, Bravaccio C (2013) Etiopathogenesis of autism spectrum disorders: fitting the pieces of the puzzle together. Med Hypotheses 81(1): 26–35

Gillberg G, Coleman M (1992) The biology of the autistic syndromes. MacKeith Press, London

Gillberg C et al (1990) Autism under age 3 years: a clinical study of 28 cases referred for autistic symptoms in infancy. J Child Psychol Psychiatry 31:921–934

Gillberg C, Gillberg IC, Steffenburg S (1992) Siblings and parents of children with autism: a controlled population-based study. Dev Med Child Neurol 34:389–398

Glasson EJ, Bower C, Petterson B, de Klerk N, Chaney G, Hallmayer JF (2004) Perinatal factors and the development of autism: a population study. Arch Gen Psychiatry 61(6):618–627

Gupta S, Aggarwal S, Rashanravan B, Lee T (1998) Th1- and Th2-like cytokines in CD4+ and CD8+ T cells in autism. J Neuroimmunol 85(1):106–109

Hallmayer J, Cleveland S, Torres A, Phillips J, Cohen B, Torigoe T et al (2011) Genetic heritability and shared environmental factors among twin pairs with autism. Arch Gen Psychiatry 68(11):1095–1102

Hobson RP (1993) Autism and the development of mind. Erlbaum, Hove

Hobson RP, Lee A (1989) Emotion-related and abstract concepts in autistic people: evidence from the British Picture Vocabulary Scale. J Autism Dev Disord 19(4):601–623

Hultman CM, Sandin S, Levine SZ, Lichtenstein P, Reichenberg A (2011) Advancing paternal age and risk of autism: new evidence from a population-based study and a meta-analysis of epidemiological studies. Mol Psychiatry 16(12):1203–1212

Jyonouchi H, Sun S, Le H (2001) Proinflammatory and regulatory cytokine production associated with innate and adaptive immune responses in children with autism spectrum disorders and developmental regression. J Neuroimmunol 120(1–2):170–179
Kilic SS, Aydin S, Kilic N, Erman F, Aydin S, Celik I (2005) Serum arylesterase and paraoxonase activity in patients with chronic hepatitis. World J Gastroenterol 11(46):7351–7354
Larsson HJ, Eaton WW, Madsen KM, Vestergaard M, Olesen AV, Agerbo E et al (2005) Risk factors for autism: perinatal factors, parental psychiatric history, and socioeconomic status. Am J Epidemiol 161(10):916–925
Lawler CP, Croen LA, Grether JK, Van de Water J (2004) Identifying environmental contributions to autism: provocative clues and false leads. Ment Retard Dev Disabil Res Rev 10(4):292–302
Lenten BJ, Wagner AC, Anantharamaiah GM, Garber DW, Fishbein MC, Adhikary L, Nayak DP, Hama S, Navab M, Fogelman AM (2002) Influenza infection promotes macrophage traffic into arteries of mice that is prevented by D-4F, an apolipoprotein A-I mimetic peptide. Circulation 106(9):1127–1132
Muhle R, Trentacoste SV, Rapin I (2004) The genetics of autism. Pediatrics 113(5):e472–e486
Narita N, Kato M, Tazoe M, Miyazaki K, Narita M, Okado N (2002) Increased monoamine concentration in the brain and blood of fetal thalidomide- and valproic acid-exposed rat: putative animal models for autism. Pediatr Res 52(4):576–579
Ozonoff S, Young GS, Carter A, Messinger D, Yirmiya N, Zwaigenbaum L et al (2011) Recurrence risk for autism spectrum disorders: a Baby Siblings Research Consortium study. Pediatrics 128(3):e488–e495
Parra S, Alonso-Villaverde C, Coll B, Ferré N, Marsillach J, Aragonès G, Mackness M, Mackness B, Masana L, Joven J, Camps J (2007) Serum paraoxonase-1 activity and concentration are influenced by human immunodeficiency virus infection. Atherosclerosis 194(1):175–181
Paucheri P, Pfanner P (1999) Il punto su psicofarmacologia nei disturbi psichiatrici dell'infanzia e dell'adolescenza. Scientific Press s.r.l
Sacco R, Militerni R, Frolli A, Bravaccio C, Gritti A, Elia M, Curatolo P, Manzi B, Trillo S, Lenti C, Saccani M, Schneider C, Melmed R, Reichelt KL, Pascucci T, Puglisi-Allegra S, Persico AM (2007) Clinical, morphological, and biochemical correlates of head circumference in autism. Biol Psychiatry 62(9):1038–1047
Sandin S, Hultman CM, Kolevzon A, Gross R, MacCabe JH, Reichenberg A (2012) Advancing maternal age is associated with increasing risk for autism: a review and meta-analysis. J Am Acad Child Adolesc Psychiatry 51(5):477–486
Shi L, Fatemi SH, Sidwell RW, Patterson PH (2003) Maternal influenza infection causes marked behavioral and pharmacological changes in the offspring. J Neurosci 23:297–302
Stagno S, Whitley RJ (1985) Herpesvirus infections of pregnancy. Part I: Cytomegalovirus and Epstein-Barr virus infections. N Engl J Med 313:1270–1274
Taft LT, Cohen HJ (1971) Hypsarrhythmia and infantile autism: a clinical report. J Autism Child Schizophr 1(3):327–336
Trevarthen C, Aitken KJ (2001) Infant intersubjectivity: research, theory, and clinical applications. J Child Psychol Psychiatry 42:3–48
Wakefield AJ, Murch SH, Anthony A, Linnell J, Casson DM, Malik M, Berelowitz M, Dhillon AP, Thomson MA, Harvey P, Valentine A, Davies SE, Walker-Smith JA (1998) Ileal-lymphoid-nodular hyperplasia, non-specific colitis, and pervasive developmental disorder in children. Lancet 351(9103):637–641

Chapter 2
Epidemiology of Autism Spectrum Disorders

Alison Presmanes Hill, Katharine Zuckerman and Eric Fombonne

Abstract In this chapter, we review existing prevalence estimates for autism spectrum disorders (ASDs) since 2000 and discuss methodological factors impacting the estimation of prevalence and the interpretation of changes in prevalence estimates over time. Possible explanations for an increase in the prevalence of ASD within and across populations are considered. Increases in ASD diagnostic rates cannot currently be attributed to a true increase in the incidence of ASD due to multiple confounding factors. It remains to be seen how changes to diagnostic criteria introduced in the *DSM-5* will impact estimates of ASD prevalence going forward.

Keywords Epidemiology · Prevalence · Autism

2.1 Introduction

Epidemiological surveys of autism were first initiated in the mid-1960s in England (Lotter 1966, 1967) and have since been conducted in over 20 countries. In this chapter, we provide a comprehensive review of the findings and methodological features of published epidemiological surveys about the prevalence of autism spectrum disorders (ASDs[1]). This chapter builds upon previous reviews (Elsabbagh et al. 2012; Fombonne 2003a, 2005; Fombonne et al. 2011; French et al. 2013; Hill et al. 2014; Williams et al. 2006) and includes the results of pertinent studies since published. The specific questions addressed are: (1) What is the range of prevalence

[1] Autism spectrum disorder (ASD) is the modern term that replaces the former pervasive developmental delay (PDD).

E. Fombonne (✉)
Department of Psychiatry, Autism Research, Institute for Development & Disability,
Oregon Health & Science University, Gaines Hall 158, Portland, OR, USA
e-mail: fombonne@ohsu.edu

A. Presmanes Hill
Research Department of Pediatrics, Oregon Health & Science University, Portland, OR, USA

K. Zuckerman
Division of General Pediatrics, Oregon Health & Science University, Portland, OR, USA

M. d. l. A. Robinson-Agramonte (ed.), *Translational Approaches to Autism Spectrum Disorder,* DOI 10.1007/978-3-319-16321-5_2

estimates for ASDs?, and (2) How should the time trends observed in the current prevalence rates of ASDs be interpreted?

2.1.1 Study Design and Methodological Issues

Epidemiologists use several measures of disease occurrence including incidence, cumulative incidence, and prevalence. Prevalence is a measure used in cross-sectional surveys (in which there is no passage of time) and reflects the proportion of subjects in a given population who suffer from the disease at that point in time. Most epidemiological studies of ASDs have assessed prevalence (point prevalence or period prevalence) as a cross-sectional approach is more appropriate for disorders where timing of diagnosis lags behind the onset of symptoms and is likely to be influenced by a range of factors unrelated to risk. In designing a prevalence study, three elements are critical: case definition, case identification (or case ascertainment), and case evaluation methods (Fombonne 2007).

2.1.1.1 Case Definition

The definition and diagnostic criteria of autism has changed over time. Starting with Kanner's definition of autism (1943), case definitions have progressively broadened to include criteria proposed by Rutter (1970), and subsequently the International Classification of Diseases, ninth revision (ICD-9; World Health Organization 1977); the *Diagnostic and Statistical Manual of Mental Disorders*, third edition (*DSM-III*; American Psychiatric Association 1980; *DSM-III-R*, American Psychiatric Association 1987), until two recent nosographies were adopted worldwide; ICD-10 (World Health Organization 1992) and the *Diagnostic and Statistical Manual of Mental Disorders*, fourth edition (*DSM-IV, American Psychiatric Association* 1994; *DSM-IV-TR*, American Psychiatric Association 2000).

Early diagnostic criteria reflected the more qualitatively severe behavioral phenotypes, usually associated with severe delays in language and cognitive skills. In the 1980s less severe forms of autism were recognized, either as a qualifier for autism occurring without intellectual disability (i.e., high-functioning autism), or as separate diagnostic categories (e.g., pervasive developmental disorders not otherwise specified [PDD-NOS], or ASD). Asperger's disorder appeared in the 1990s, with unclear validity, particularly with respect to its differentiation from high-functioning autism. Some ASD subtypes that were described in *DSM-III* subsequently disappeared (e.g., Autism-Residual State); however, other nomenclatures have since added new diagnostic categories, such as "atypical autism" and "PDD unspecified" (ICD-10).

The changes now occurring with the introduction of *Diagnostic and Statistical Manual of Mental Disorders*, fifth edition (*DSM-5*; American Psychiatric Association 2013), may impact prevalence estimates in the future. *DSM-5* proposes a single new category of ASDs, conceptually equivalent to the previous diagnostic class of

PDDs. However, fewer diagnostic criteria have been retained that are combined in two clusters of social communication deficits, and restricted patterns of behaviors and interests. The removal of the loosely defined PDD-NOS that was in *DSM-IV-TR* (American Psychiatric Association 2000) will likely increase the specificity of the ASD diagnostic category, and the removal of Asperger disorder as a separate category is consistent with the research that has generally failed to provide evidence for the discriminant validity of this diagnostic concept vis-à-vis the forms of autistic disorder that are not associated with severe language impairments or intellectual deficits.

The impact of *DSM-5* changes remains to be fully assessed in the context of epidemiological surveys. Two recent population-based surveys have addressed this issue. Maenner and colleagues (2014) retrospectively applied the new diagnostic criteria to a previously obtained population-based sample from the Centers for Disease Control and Prevention (CDC) 2006 and 2008 surveillance years. They found that 81.2% of children classified as having ASD according to DSM-IV-TR (American Psychiatric Association 2000) also met *DSM-5* criteria (American Psychiatric Association 2013), resulting in a *DSM-5* based prevalence of 100/10,000—an estimate lower than the CDC 2006 and 2008 estimates. In addition, 304 children met *DSM-5* but not *DSM-IV-TR*. In a similar study, Kim and colleagues (2014) reported that 92% of children with ASD according to *DSM-IV-TR* also met *DSM-5* criteria. However, when *DSM-5* ASD and social communication disorder (SCD; a new diagnostic category in *DSM-5*) were considered together, there was no significant change in the prevalence estimate (Kim et al. 2014). It is important to note that new diagnostic information required in *DSM-5* (e.g., emphasis on sensory processing deficits) is generally not available in prior studies, leading to potentially biased estimates. Additionally, previous studies are often constrained in sampling children with a *DSM-IV* PDD diagnosis and cannot therefore accurately estimate the proportion of children who did not meet the criteria for *DSM-IV* yet would have met those for *DSM-5*.

While there is currently high reliability overall regarding diagnosis of ASDs and commonality of concepts across experts, differences still persist between nomenclatures about the terminology and operationalized criteria of ASDs. It is unclear to what extent the changing nomenclature of ASDs plays a role in prevalence estimates described in epidemiological studies. More studies are on their way that will provide further examination of the impact on prevalence estimates of narrowing the ASD definition in *DSM-5*.

2.1.1.2 Case Identification/Ascertainment

When a population is identified for a survey, different strategies are employed to find individuals matching the study's case definition. Some studies rely solely on service provider databases (Chien et al. 2011; Croen et al. 2002; Davidovitch et al. 2013), special education databases (Fombonne et al. 2006; Gurney et al. 2003; Lazoff et al. 2010; Maenner and Durkin 2010), or national registers (Al-Farsi et al. 2011; Parner et al. 2012; Samadi et al. 2011) for case identification. These studies

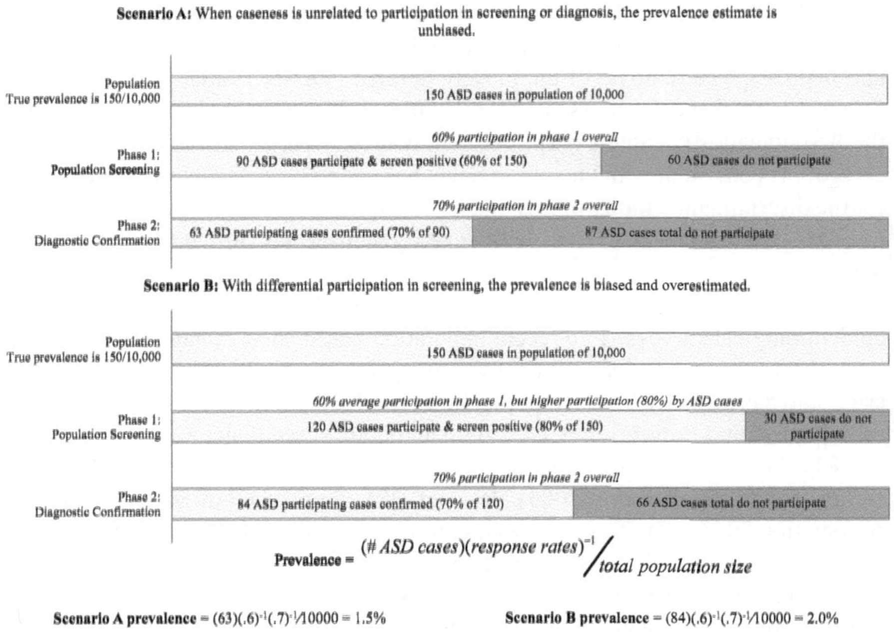

Fig. 2.1 Assuming a true ASD prevalence of 150/10,000 and a sensitivity of 100% for the screening process and total accuracy in the diagnostic confirmation, weighting back phase 2 data results in an unbiased prevalence estimate when caseness is unrelated to participation in screening (Scenario A), but when participation in screening is more likely for ASD cases than for non-cases (Scenario B), prevalence will be overestimated (see discussion in text)

have the common limitation of relying on a population group that was readily accessible, rather than sampling from the population at large. As a result, individuals with the disorder who are not in contact with services are not included as cases, leading to an underestimation of prevalence. This limitation is particularly problematic in communities with recognized limitations in available services.

Other investigations have relied on a multistage approach to identify cases in underlying populations (Centers for Disease Control and Prevention 2012; Idring et al. 2012; Kim et al. 2011). In these studies' first screening stage, a wide net is cast to identify subjects possibly affected with ASD, with the final diagnostic status being determined at subsequent stages. This process often consists of sending letters or screeners to school and health professionals, searching for the possible cases of autism. Few such investigations rely on systematic sampling techniques that would ensure a near complete coverage of the target population, and screening often varies substantially in ascertainment of all relevant data sources. Additionally, surveyed areas often differ in terms of specific educational or health care systems available, and inclusion information sent often varies in reliability and validity. Finally, uneven participation rates in the screening stage can lead to variation in the screening efficiency of surveys.

To illustrate how differential participation in the screening stage affect prevalence estimates, two hypothetical scenarios are illustrated in Fig. 2.1, both of which are based on a true ASD prevalence of 150/10,000 and a sensitivity of 100% for the screening process and total accuracy in the diagnostic confirmation. In Scenario A, we assume 60 % participation for ASD and non-ASD cases in the first screening stage, resulting in 90 participating ASD cases that screen positive. With 70 % participation for both ASD and non-ASD cases in the diagnostic stage, we would identify and confirm 63 ASD cases in the second phase. Weighting back phase 2 data, we would obtain an unbiased prevalence estimate of 1.5 % (or 150/10,000) in this scenario. In Scenario B, we also assume 60 % overall participation, but with an 80 % participation rate for ASD cases, reflecting a scenario in which individuals with ASD are more likely to participate in the first screening stage than non-ASD cases. Thus, with the same participation rates in the first screening (60 %) and final diagnostic stages (70 %), we identify 84 ASD cases and calculate a biased prevalence estimate of 2 % (200/10,000), an estimate that is 0.5 % higher than true prevalence. The bias arises for two reasons: (1) participation in screening is associated with case status (here, with ASD cases more likely to participate than non-cases); and (2) as investigators typically have no such information, weights used for prevalence estimation were not adjusted correspondingly, resulting in the upward bias.

It is also possible that individuals with ASD participate less than non-cases, which would result in underestimates of prevalence. For example, Posserud and colleagues (2010) reported the ASD prevalence of 72/10,000 in their identified sample and estimated a prevalence of 128/10,000 in no responders (based on teacher ratings during the screening phase), indicating increased refusal rates among those with more ASD symptoms. Unfortunately, few studies have been able to estimate the extent to which willingness or refusal to participate is associated with final caseness, so it is not known what effect differential participation rates at different phases in population surveys may have on prevalence estimates.

The sensitivity of the screening methodology is difficult to gauge in autism surveys, as the proportion of children truly affected with the disorder but not identified in the screening stage (false negatives) remains generally unmeasured. Few studies provided an estimate of the reliability of the screening procedure. The usual approach, which consists of randomly sampling screen-negative subjects to adjust estimates, has not been generally used, mainly due to the relatively low frequency of ASD, which makes such a strategy both imprecise and costly.

As an example, the surveys conducted by US CDC (2007a, 2007b, 2009, 2012, 2014) rely, for case ascertainment, on scrutinizing educational and medical records. Children not accessing such services cannot be identified. Although some recent surveys that systematically screen the normal school population might detect a large pool of unidentified cases (Kim et al. 2011), it remains to be seen if this applies to most populations and requires change in sampling approaches for surveying autism. Of note, the CDC methodology identifies ASD cases without prior official ASD diagnosis (21 % of identified cases in 2008; Centers for Disease Control and Prevention 2012), suggesting that underidentification is a widespread phenomenon.

Since more recent prevalence studies suggest that autism can no longer be regarded as rare, screening for false negatives may become a more common strategy. Currently, however, prevalence estimates must be understood as underestimates of "true" prevalence rates, with the magnitude of this underestimation unknown in each survey.

2.1.1.3 Case Evaluation

When the screening phase is completed, subjects identified as positive go through a more in-depth diagnostic evaluation to confirm case status. Similar considerations about methodological variability across studies apply in more intensive assessment phases. The information used to determine diagnosis usually involves a combination of data from informants (parents, teachers, pediatricians, other health professionals, etc.) and data sources (medical records, educational sources), with a direct assessment of the person with autism being offered in some but not all studies. When subjects are directly examined, assessments typically use various diagnostic instruments, ranging from a typical unstructured examination by a clinical expert (but without demonstrated psychometric properties) to the use of batteries of standardized measures by trained research staff. The Autism Diagnostic Interview-Revised (ADI-R; Lord et al. 1994), and/or the Autism Diagnostic Observation Schedule (ADOS; Lord et al. 2000) have been increasingly used in the most recent surveys (Table 2.1).

Obviously, surveys of large populations, such as those conducted by the CDC (2007a, 2007b, 2009, 2012) or in national registers (Idring et al. 2012), cannot include direct diagnostic assessment of all subjects by researchers. However, investigators generally improve the accuracy of caseness determinations by undertaking, on a randomly selected subsample, a more complete diagnostic workup (Rice et al. 2007). The CDC surveys have established a methodology for surveys of large populations based on the screening of the population using multiple data sources, standardized records abstraction, and systematic review and scoring of the data gathered in the screening phase. In the less obvious cases, this information is combined with input from experienced clinicians with known reliability and validity. This methodology is adequate for large samples, and is likely to be used in the future for surveillance efforts.

2.2 Systematic Review of Prevalence Estimates

2.2.1 Search Strategies

Keeping in mind the range and limitations of case definition, identification, and evaluation methods employed in epidemiological surveys, we present the results of epidemiological reports conducted since 2000 in Table 2.1. These reports were identified from previous reviews of epidemiological surveys (Elsabbagh et al. 2012; Fombonne 2003b, 2003a, 2005, 2009a; Fombonne et al. 2011; French et al. 2013; Williams et al. 2006) and through systematic searches using major scientific literature databases (Medline, PsycINFO, Embase, PubMed). Where multiple surveys based on the same or overlapping populations were evident, the publication listed is the most detailed and comprehensive account. For example, surveys conducted by

Table 2.1 Prevalence surveys of ASDs since 2000.

Year	Authors	Country	Area	Population	Age	Number Affected	Diagnostic Criteria	% with Normal IQ	Gender Ratio (M:F)	Prevalence /10,000	95% CI
2000	Baird et al.	UK	South East Thames	16,235	7	94	ICD-10	60	15.7 (83:11)	57.9	46.8; 70.9
2000	Powell et al.	UK	West Midlands	58,654[a]	1–5	122	Clinical, ICD-10, DSM-IV	–	–	20.8	17.3; 24.9
2001	Bertrand et al.	USA	New Jersey	8896	3–10	60	DSM-IV	51	2.7 (44:16)	67.4	51.5; 86.7
2001	Chakrab-arti and Fombonne	UK	Stafford	15,500	2.5–6.5	96	ICD-10	74.2	3.8 (77:20)	61.9	50.2; 75.6
2001	Fombonne et al.	UK	England and Wales	10,438	5–15	27	DSM-IV, ICD-10	55.5	8.0 (24:3)	26.1	16.2; 36.0
2002	Scott et al.	UK	Cambridge	33,598	5–11	196	ICD-10	–	4.0 (–)	58.3[a]	50.7; 67.1[a]
2003	Yeargin-All-sopp et al.	USA	Atlanta, GA	289,456	3–10	987	DSM-IV	31.8	4.0 (787:197)	34.0	32; 36
2003	Gurney et al.	USA	Minnesota (2001–2002)	787,308[a]	6–11	4094	Receipt of MN special educa-tion services	–	–	52.0[a]	50.4; 53.6[a]
2003	Lingam et al.	UK	North East London	186,206	5–14	567	ICD-10	–	4.8 (469 :98)	30.5[a]	27.9; 32.9[a]
2004	Icasiano et al.	Australia	Barwon	45,153[a]	2–17	177	DSM-IV	53.4	8.3 (158:19)	39.2	33.8; 45.4[a]
2005	Chakrab-arti and Fombonne	UK	Stafford	10,903	4–6	64	ICD-10	70.2	6.1 (55:9)	58.7	45.2; 74.9

Table 2.1 (continued)

Year	Authors	Country	Area	Population	Age	Number Affected	Diagnostic Criteria	% with Normal IQ	Gender Ratio (M:F)	Prevalence /10,000	95% CI
2006	Baird et al.	UK	South Thames (1990–1991)	56,946	9–10	158	ICD-10	45	3.3 (121:37)	116.1	90.4; 141.8
2006	Fombonne et al.	Canada	Montreal	27,749	5–17	180	DSM-IV	–	4.8 (149:31)	64.9	55.8; 75.0
2006	Harrison et al.	UK	Scotland	134,661	0–15	443b	ICD-10, DSM-IV	–	7.0 (369:53)	44.2	39.5; 48.9
2006	Gillberg et al.	Sweden	Göteborg	32,568	7–12	262	DSM-III, DSM-IV, Gill-berg's criteria for AS	–	3.6 (205:57)	80.4	71.3; 90.3
2006	Ouellette-Kuntz et al.	Canada	Manitoba and Prince Edward Island	227,526	1–14	657	DSM-IV	–	4.1 (527:130)	28.9[a]	26.8; 31.2[a]
2007	Croen et al.	USA	Northern California (1995–1999)	132,844	5–10	593	ICD-9-CM	–	5.5 (501:92)	45	41.2; 48.4[a]
2007[b]	CDC	USA	6 states	187,761	8	1252	DSM-IV-TR	38 to 60[d]	2.8 to 5.5	67.0	63.1; 70.5[a]
2007[c]	CDC	USA	14 states	407,578	8	2685	DSM-IV-TR	55.4[e]	3.4 to 6.5	66.0	63; 68
2007	Latif and Williams	UK	South Wales	39,220	0–17	240	ICD-10, DSM-IV, Kanner's and Gillberg's criteria	–	6.8 (–)	61.2	53.9; 69.4[a]

Table 2.1 (continued)

Year	Authors	Country	Area	Population	Age	Number Affected	Diagnostic Criteria	% with Normal IQ	Gender Ratio (M:F)	Prevalence /10,000	95% CI
2008	Wong and Hui	China	Hong Kong Registry	4,247,206	0–14	682	DSM-IV	30	6.6 (592:90)	16.1	14.9; 17.3[a]
2008	Montiel-Nava and Pena	Venezuela	Maracaibo	254,905	3–9	430	DSM-IV-TR	–	3.3 (329:101)	17	13; 20
2008	Kawamura et al.	Japan	Toyota	12,589	5–8	228	DSM-IV	66.4	2.8 (168:60)	181.1	159.2; 205.9[a]
2008	Williams et al.	UK	Avon	14,062	11	86	ICD-10	85.3	6.8 (75:11)	61.9	48.8; 74.9
2009	Baron-Cohen et al.	UK	Cambridgeshire	8824	5-9	83	ICD-10	–	–	94[f]	75; 116
2009	Nicholas et al.	USA	South Carolina	8156	4	65	DSM-IV-TR	44.2	4.7	80	61; 99
2009	van Balkom et al.	Netherlands	Aruba	13,109	0–13	69	DSM-IV	58.8	6.7 (60:9)	52.6	41.0; 66.6
2009	CDC	USA	11 states	308,038	8	2,757	DSM-IV-TR	59	4.5	90	86; 93
2010	Fernell and Gillberg	Sweden	Stockholm	24,084	6	147	DSM-IV, DSM-IV-TR, ICD-10	33	5.1 (123:24)	62	52; 72
2010	Lazoff et al.	Canada	Montreal	23,635	5–17	187	DSM-IV	–	5.4 (158:29)	79.1	67.8; 90.4
2010	Barnevik-Olsson et al.	Sweden	Stockholm	113,391	6–10	250	DSM-IV	0	–	22	19.4; 25.0[a]

Table 2.1 (continued)

Year	Authors	Country	Area	Population	Age	Number Affected	Diagnostic Criteria	% with Normal IQ	Gender Ratio (M:F)	Prevalence /10,000	95% CI
2010	Maenner and Durkin	USA	Wisconsin	428,030	Elementary school–aged	3831	DSM-IV like criteria for WI special education services (by school district)	–	–	90	86.7; 92.4[a]
2010	Posserud et al.	Norway	Bergen	9,430	7–9	16	DSM-IV, ICD-10 Included DAWBA and DISCO	–	7 (14:2)	87[g]	–
2011	Al-Farsi et al.	Oman	National Register	528,335	0–14	113	DSM-IV-TR	–	2.9 (84:29)	1.4	1.2; 1.7
2011	Brugha et al.	UK	England	7333	16–98	72	ADOS	100	3.8	98.2	30; 165
2011	Kim et al.	S. Korea	Goyang City	55,266	7–12	201	DSM-IV	31.5	3.8	264	191; 337
2011	Mattila et al.	Finland	Northern Ostrobothnia County	5484	8	37	DSM-IV-TR included ADOS-G and ADI-R	65	1.8	84	61; 115
2011	Parner et al.[h]	Australia	Western Australia (1994–1999)	152,060	0–10	678	DSM-IV, DSM-IV-TR	–	4.1	51	47; 55.3
2011	Samadi et al.	Iran	National Register	1,320,334	5	826	ADI-R	–	4.3	6.4	5.84; 6.70

Table 2.1 (continued)

Year	Authors	Country	Area	Population	Age	Number Affected	Diagnostic Criteria	% with Normal IQ	Gender Ratio (M:F)	Prevalence /10,000	95% CI
2011	Chien et al.	Taiwan	National Health Research Institute	229,457[a]	0–18	659	ICD-9	–	3.7	28.7	26.6; 31[a]
2011	Windham et al.[i]	USA	San Francisco Bay [A-Za-z_-'&;]{3,20} (1994,1996)	80,249	9	374	"Full syndrome autism"—CA Dept. of Developmental Services, receipt of CA special education services, or DSM-IV	–	6.2 (324:50)	47	42; 52
2012	CDC	USA	14 states	337,093	8	3820	DSM-IV-TR	38	4.6	113	110; 117
2012	Idring et al.	Sweden	Stockholm County Register	444,154	0–17	5100	ICD-09, ICD-10, DSM-IV	57.4	2.6	115	112; 118
2012	Isaksen et al.	Norway	Oppland and Hedmark	31,015	6–12	158	ICD-10 included ADOS-G and ADI-R	–	4.27 (128:30)	51	43; 59
2012	Kočovská, Biskuptso, et al.[j]	Denmark	Faroe Islands	7128	15–24	67	ICD-10, DSM-IV, Gillberg's criteria	–	2.7[a] (49:18)	94	73; 119
2012	Nygren et al.	Sweden	Göteborg	5007	2	40	DSM-IV-TR	63[a]	4 (32:8)	80	57; 109

Table 2.1 (continued)

Year	Authors	Country	Area	Population	Age	Number Affected	Diagnostic Criteria	% with Normal IQ	Gender Ratio (M:F)	Prevalence /10,000	95% CI
2012	Parner et al.[k]	Denmark	National Register (1980–2003)	1,311,736	6–29	9556	ICD-8, ICD-9, ICD-10	–	4.1	72.9[a]	71.4; 74.3[a]
2013	Davidovitch et al.	Israel	Maccabi HMO Registry	423,524	1–12	2034	DSM-IV	–	5.2	48	45.9; 50.1
2013	Ouellette-Kuntz et al.[l]	Canada	Prince Edward Island & Southeastern Ontario (2010)	89,786	2–14	1173	Diagnosis of ASD from qualified professional - National Epidemiologic Database for the Study of Autism in Canada		4.8[a] (896:186)	130.6[a]	123.4; 138.3[a]
2013	Saemundsen et al.	Iceland	National Database	22,229	6	267	ICD-10 included ADOS & ADI-R	54.7	2.8 (197:70)	120.1	106.6; 135.3
2013	Taylor et al.[m]	UK	National Database	256,278	8	616	DSM-IV according to General Practice Research Database	–	5.1[a] (515:101)	24.0[a]	22.2; 26.0[a]
2014	CDC	USA	11 states	363,749	8	5338	DSM-IV-TR	69[d]	4.5	147	142.9; 150.7

Table 2.1 (continued)

Year	Authors	Country	Area	Population	Age	Number Affected	Diagnostic Criteria	% with Normal IQ	Gender Ratio (M:F)	Prevalence /10,000	95% CI
2014	Atladottir et al.	USA	Denmark, Finland, Sweden, Western Australia	4,534,236	4–20	35,863	ICD-8, ICD-9, ICD-10, DSM-IV	–	–	79.1[a]	78.2; 79.9[a]

[a] Calculated by the authors
[b] This is the prevalence for children aged 6–11 in the 2001–2002 school year
[c] Estimated using a capture-recapture analysis, the number of cases used to calculate prevalence was estimated to be 596
[d] Specific values for % with normal IQ and confidence intervals are available for each state prevalence
[e] Average across seven states
[f] Rate based on special education needs register. A figure of 99/10,000 is provided from a parental and diagnostic survey. Other estimates in this study vary from 47 to 165/10,000 deriving from various assumptions made by the authors
[g] This was the prevalence estimate based on the identified sample; when adjusted for nonresponders, the prevalence was estimated to be even higher (87/10,000)
[h] Note that this is an updated prevalence estimate: previous estimates have been reported by Nassar et al. (2009; birth years: 1983–1999; prevalence: 23.4/10,000) and Leonard et al. (2011; birth years: 1984–1999; singletons; prevalence: 30/10,000) using the same register in Western Australia
[i] Data for 1996 birth cohort; overall prevalence for both 1994 and 1996 cohorts was 47/10000 although other specific values differed slightly. This study population may overlap to some degree with Croen et al. (2007), where 1995–1999 births only at Kaiser Permanente Northern California (KPNC) were examined; KPNC was one of three types of health-based sources used in Windham et al. (2011)
[j] Note that this is an updated prevalence estimate: a previous estimate of 53.3/10,000 was reported by Ellefsen et al. (2007) based on a survey of the same geographical area with the same cohort
[k] Note that this is an updated prevalence estimate: a previous estimate was reported by Lauritsen et al. (2004; birth years: 1971–2000; prevalence: 34.4/10,000) and Parner et al. (2011; birth years: 1994–1999; prevalence: 68.5/10,000) using the same national register in Denmark
[l] Prevalence estimate calculated by the authors based on combined data from two regions in 2010
[m] Prevalence estimate calculated by the authors based on combined data from boys and girls

the CDC (2007a, 2007b, 2009, 2012, 2014) as part of the autism and developmental disabilities monitoring (ADDM) network are each included in the table, although additional accounts for individual states are available elsewhere (Nicholas et al. 2008; Pinborough-Zimmerman et al. 2012; Rice et al. 2010; Zahorodny et al. 2014).

2.2.2 Inclusion and Exclusion Criteria

The following criteria were set to select epidemiological surveys included in Table 2.1:

- The full article was published in English.
- The minimum population was 5000.
- The survey included independent validation of caseness by professionals. In addition, surveys that imposed further non-ASD criteria were excluded.
- The following information categories were included or could be ascertained based on information from the survey: country and area where the survey was conducted, size of the population for which the prevalence estimate was ascertained, age range of participants, number of children affected, diagnostic criteria used in case definition, and prevalence estimate (number per 10,000). Where available, we also report the proportion of subjects with intelligence quotient (IQ) within the normal range and gender ratios.

2.2.3 Prevalence Estimates for Combined ASDs since 2000

The results of the 53 surveys that estimated the prevalence of the whole spectrum of ASDs are summarized in Table 2.1. All selected surveys were published since 2000, with the majority (55 %) published in 2009 or later. The studies were performed in 18 different countries (including 14 in the UK and 12 in the USA, of which 5 were conducted by the CDC). Sample sizes ranged from 5007 to 4.5 million (median: 58,654; mean: 346,776). Ages of the surveyed populations ranged from 0 to 98 (median: 8; mean: 9). One study was specifically conducted on adults and provided the only estimate (98.2/10,000) thus far available for adults (Brugha et al. 2011). Two surveys focusing on toddlers (Nygren et al. 2012) and preschoolers (Nicholas et al. 2009) provided estimates of approximately 80 per 10,000. In the 50 remaining surveys, the average median age was 8.23 years ($SD = 2.8$).

The diagnostic criteria used in 53 studies reflected the reliance on modern diagnostic schemes (11 studies used ICD-10, 25 the *DSM-III*, *DSM-IV*, or *DSM-IV-TR*; both schemes being used simultaneously in 9 studies). Assessments were often performed with standardized diagnostic measures (i.e., ADI-R and ADOS). In 26 studies where IQ measures were reported, the proportion of subjects within the normal IQ range varied from 0 to 100 % (median: 55.4 %; mean: 53.9 %), a proportion that reflects the lesser association, or lack thereof, between intellectual impairment and

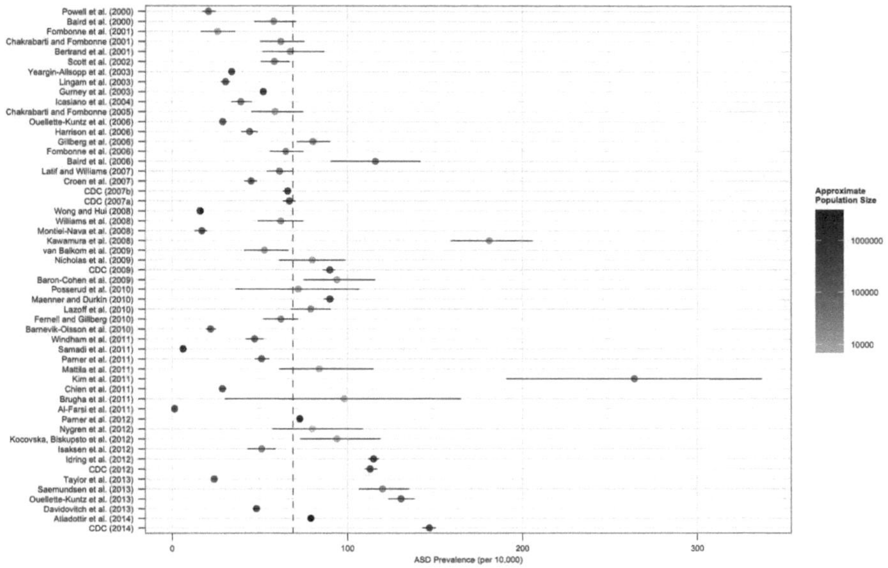

Fig. 2.2 Prevalence estimates for ASD since 2000 (per 10,000 with 95% confidence intervals; also see Table 2.1). The *dashed vertical line* denotes the mean prevalence of 69/10,000 across all 53 surveys

milder forms of ASDs. Overrepresentation of males was seen in the 47 studies reporting gender ratios, with male/female ratio ranging from 1.8:1 to 15.7:1 (median: 4.5:1; mean: 4.9:1).

There was a 189-fold variation in ASD prevalence, ranging from 1.4/10,000 to 264/10,000 (see Fig. 2.2). There was also substantial variation in confidence interval width, reflecting variation in sample sizes and consequently in each study's precision (range: 0.5–146; mean interval width: 22.4). However, some consistency in ASD prevalence is found in the center of this distribution, with a median rate of 61.9/10,000 and a mean rate of 68.9/10,000 (interquartile range: 44.2–84.0/10,000). Prevalence was negatively associated with sample size (Kendall's tau: −0.23, $p=0.01$), with small-scale studies reporting higher prevalence.

There was also a significant positive correlation between ASD prevalence estimates and publication year (Kendall's tau: 0.26, $p=0.007$), with higher rates in more recent surveys. Eight studies since 2000 reported ASD prevalence estimates higher than 100/10,000 (Baird et al. 2006; Centers for Disease Control and Prevention 2012; Idring et al. 2012; Kawamura et al. 2008; Kim et al. 2011; Ouellette-Kuntz et al. 2006; Saemundsen et al. 2013). Baird et al. (2006) and Kim et al. (2011) both employed proactive case finding techniques, relying on multiple and repeated screening phases, involving both different informants at each phase and surveying the same cohorts at different ages, which certainly enhanced the sensitivity of case identification. Multisource active surveillance techniques, as employed in the Stockholm Youth Cohort (Idring et al. 2012) and by the CDC's ADDM Network (2012, 2014), also improve identification of individuals with ASD. The most recent

CDC prevalence estimate of 147 per 10,000 reflects the highest estimate to date across all of the previous ADDM Network reports (Centers for Disease Control and Prevention 2014).

Overall, results of recent surveys agree that an average figure of 69/10,000 can be used as the current estimate for the spectrum of ASDs. The convergence of estimates around 60 to 90 per 10,000 for all ASDs combined, conducted in different regions and countries by different teams, is striking especially when derived from studies with improved methodology. The prevalence figure of 69/10,000 (equivalent to 6.9/1000 or.69%) translates into 1 child out of 145 with an ASD diagnosis. This estimate is now the best current estimate for the ASD prevalence. However, it represents an average and conservative figure, and substantial variability exists between studies and within studies, across sites or areas.

2.3 Time Trends in Prevalence and Their Interpretation

The debate on the hypothesis of a secular increase in rates of autism has been obscured by a lack of clarity in the measures of disease occurrence. As noted previously, it is crucial to differentiate prevalence from incidence, since only incidence rates can be used for causal research, and prevalence and incidence will increase when case definition is broadened or case ascertainment is improved. Moreover, epidemiological surveys of ASDs possess unique design features that could account almost entirely for between-study variation in prevalence estimates, making time trends even more difficult to gauge. Time trends in prevalence estimates can therefore only be evaluated in investigations that hold methodological parameters under strict control over time. Such requirements must be considered when reviewing evidence for a secular increase in rates of ASDs, or testing for the "epidemic" hypothesis.

The epidemic hypothesis emerged in the 1990s when, in most countries, increasing numbers were diagnosed with ASDs leading to an upward trend in children registered in service providers' databases that was paralleled by higher prevalence rates in epidemiological surveys. These trends were interpreted as evidence that the actual population incidence of ASDs was increasing. However, because methodological factors contribute to variability in prevalence estimates, these must be considered before concluding that there is a true rise in the number of children diagnosed with ASDs and include the following:

2.3.1 Use of Referral Statistics

Increasing numbers of children referred to specialist services or known to special education registers have been taken as evidence for increased ASD incidence. Such upward trends have been seen in many different countries (Gurney et al. 2003;

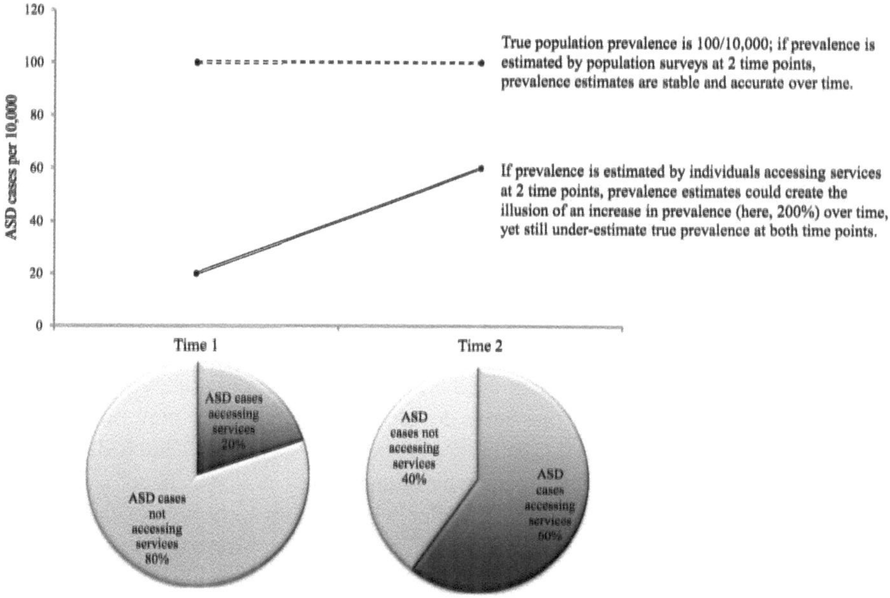

Fig. 2.3 Assuming a constant incidence and prevalence of 100/10,000 between Time 1 and Time 2 (meaning there is no "epidemic"), prevalence estimates that rely solely on service access counts not only underestimate the true prevalence but may also create the illusion of rising prevalence over time (see discussion in text)

Lotter 1966; Shattuck 2006; Taylor et al. 1999), all occurring in the late 1980s and early 1990s. However, trends over time in *referred* samples are confounded by referral patterns, availability of services, heightened public awareness, decreasing age at diagnosis, and changes over time in diagnostic concepts and practices.

As an illustration, Fig. 2.3 contrasts two methods for surveying ASD using hypothetical data: one based on sampling from the total population, and the other relying solely on service access counts. Here, assuming a constant incidence and prevalence of 100/10,000 between Time 1 and Time 2 (meaning there is no epidemic), population surveys at two time points result in prevalence estimates that are not only accurate but also stable over time, showing no prevalence change in the target population. However, if prevalence is estimated based only on service access counts where the number of ASD individuals accessing services increases from 20 to 60 % over time, prevalence would be underestimated at both time points, yet would appear to rise 200 % while the underlying true incidence and prevalence remained stable. Such a pattern of results was recently reported based on special education data in Wisconsin (Maenner and Durkin 2010), in which ASD prevalence rates were stable between 2002 and 2008 in school districts with initially high baseline prevalence rates (≈ 120/10,000), whereas school districts with low baseline rates experienced significant increases in prevalence (e.g., in one district rates rose from 5 to 70/10,000; corresponding to a 1300 % increase in 6 years). Failure to control for these confounding factors was obvious in previous reports (Fombonne

2001), including widely quoted reports from California Developmental Database Services (California Department of Developmental Services 2003).

Additionally, the decreasing age at diagnosis results in itself to increasing numbers of young children being identified in official statistics (Wazana et al. 2007) or referred to specialist medical and educational services. Earlier identification of children from the prevalence pool may therefore result in increased service activity that may lead to a misperception by professionals of an epidemic.

2.3.2 Diagnostic Substitution

Another possible explanation for increased prevalence in a diagnostic category is that children presenting with the same developmental disability may receive one particular diagnosis initially and another diagnosis subsequently. Such diagnostic substitution (or switching) may occur when diagnostic categories become increasingly familiar to health professionals and/or when access to better services is ensured by using a new diagnostic category.

The strongest evidence of diagnostic substitution contributing to ASD prevalence increase was shown in a complex analysis of Department of Education Data in 50 US states (Shattuck 2006), indicating that a relatively high proportion of children previously diagnosed with mental retardation (MR) were subsequently identified as having ASD. Shattuck showed that the odds of having ASD increased by 1.21 during 1994–2003 while the odds of having learning disability (LD) (odds ratio [OR] = 0.98) and MR (OR = 0.97) decreased. Shattuck (2006) further demonstrated that the growing ASD prevalence was directly associated with decreasing prevalence of LD and MR within states, and that a significant downward deflection in the historical trajectories of LD and MR occurred when ASD became reported in the USA as an independent category in 1993–1994.

Using individual level data, a newer study reexamined the hypothesis of diagnostic substitution in the California department of developmental services dataset (King and Bearman 2009) and showed that 24 % of the increase in caseload was attributable to diagnostic substitution (from MR to ASD). It is important to keep in mind that other types of diagnostic substitution are likely to have occurred as well for milder forms of ASD. For example, children currently diagnosed with Asperger's disorder may be previously diagnosed with other psychiatric diagnoses (i.e., obsessive-compulsive disorder, school phobia, social anxiety, etc.) in clinical settings before the developmental nature of their condition was fully recognized (Fombonne 2009b).

2.3.3 Cross-Sectional Variability in Epidemiological Surveys

Evidence that method factors could account for most of the variability in published prevalence estimates comes from a direct comparison of eight recent surveys

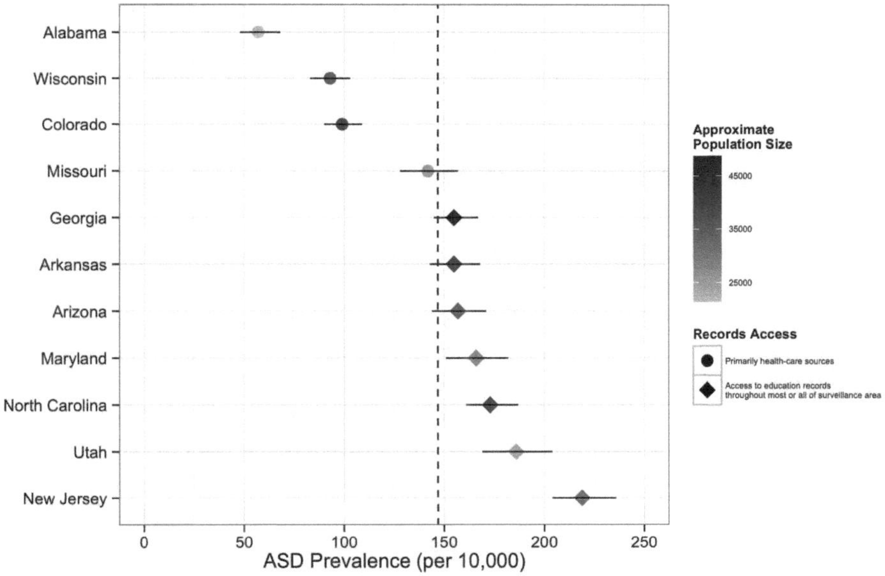

Fig. 2.4 Estimated prevalence of ASDs (with 95 % confidence intervals) among children aged 8 years in the USA by ADDM site and type of records access in the 2010 surveillance year (CDC, 2014). The *dashed vertical line* denotes the average prevalence estimate of 147/10,000 across all sites

conducted in the UK and the USA (Fombonne 2005). In each country, four surveys were conducted around the same year and with similar age groups. As there is no reason to expect large variations in between-area differences in rates, prevalence estimates should therefore be comparable within each country. However, there was a 6-fold variation in rates for the UK surveys, and a 14-fold variation in the US rates. In each set of studies, high rates were found when intensive population-based screening techniques were employed, whereas lower rates were found in studies relying on passive administrative methods for case finding. Since no passage of time was involved, the magnitude of these gradients in rates is likely to reflect methodological differences.

Even more convincing evidence comes from the most recent survey by the CDC on 363,749 children aged 8 in 2010, where an average prevalence of 147/10,000 was reported across 11 US states (Centers for Disease Control and Prevention 2014). One striking finding in this report is the almost four-fold variation in prevalence rates by state (range: 57–219 per 10,000; see Fig. 2.4). Across individual states, Alabama had the lowest rate of 57/10,000, whereas New Jersey had the highest rate of 219/10,000 (Centers for Disease Control and Prevention 2014). Estimated ASD prevalence was significantly lower in states that had access to health data sources only compared to that of the states where educational data was also available (97.7 versus 149 out of 10,000, respectively), a factor that is consistently associated with higher prevalence rates in the ADDM Network. It would be surprising if there were truly this much inherent state-to-state variability in the number of children with autism in the USA. Thus, these differences likely reflect ascertainment variability across sites in a study that was otherwise performed with the same methods, at the same time, on children of the same age, and within the same country.

2.3.4 Repeated Surveys in Defined Geographical Areas

Repeated surveys, using the same methodology and conducted in the same geographical area at different time-points, can potentially yield useful information on time trends if methods are kept relatively constant. The Göteborg studies (Gillberg 1984; Gillberg et al. 1991) provided three prevalence estimates that increased over a short period of time from 4.0 (1980) to 6.6 (1984) to 9.5/10,000 (1988), the gradient being even steeper in urban areas only (Gillberg et al. 1991). However, comparison of these rates is not straightforward, as different age groups were included in each survey. Furthermore, increased prevalence was associated with improved detection among those with intellectual delays in the second survey, and with improved detection of cases born to immigrant parents in the third survey, suggesting that migration into the area could be a key explanation. Taken in conjunction with a change in local services and a progressive broadening of the autism definition over time (Gillberg et al. 1991), findings provide weak evidence for increased autism incidence. Similarly, studies conducted in Japan at different points in time in Toyota (Kawamura et al. 2008) and Yokohama (Honda et al. 2005; Honda et al. 1996) showed rises in prevalence rates that their authors interpreted as reflecting the effect of both improved population screening of preschoolers and a broadening of diagnostic concepts and criteria.

Two separate surveys of children born between 1992 and 1995, and between 1996 and 1998 in Staffordshire, UK (Chakrabarti and Fombonne 2001, 2005), were performed with rigorously identical methods for case definition and case identification. The prevalence for combined ASDs was comparable and not statistically different in the two surveys (Chakrabarti and Fombonne 2005), suggesting no upward trend in overall rates of ASDs, at least during the short time interval between studies.

2.3.5 Birth Cohorts

In large surveys encompassing wide age ranges, increasing prevalence among most recent birth cohorts could be interpreted as indicating a secular increase in ASD incidence, provided that alternative explanations can be confidently eliminated. This analysis was used in two large French surveys (Fombonne and Du Mazaubrun 1992; Fombonne et al. 1997). The surveys included birth cohorts from 1972 to 1985 (735,000 children, 389 of whom had autism). When pooling the data of both surveys, age-specific rates showed no upward trend (Fombonne et al. 1997).

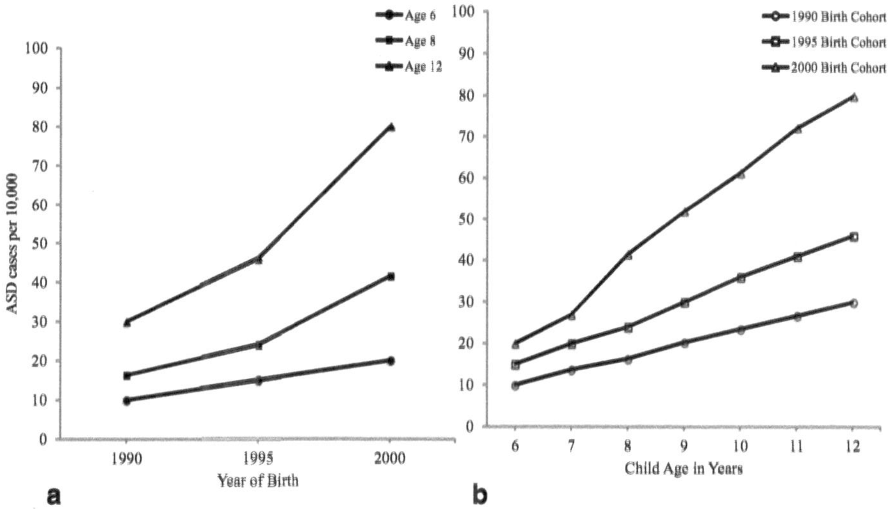

Fig. 2.5 Using hypothetical data, Figure 5a illustrates rising prevalence rates among 6-, 8-, and 12-year-old children across three different birth cohorts. Prevalence rates also increase within birth cohorts as they age (Figure 5b), potentially coinciding with changes in patterns of referral, service availability, public awareness, and diagnostic concepts and practices (see discussion in text)

However, data assessing birth cohorts can be problematic, as illustrated in Fig. 2.5, which shows an increase in the prevalence of ASD by year of birth across three hypothetical successive birth cohorts (a cohort effect; Fig. 2.5a). Within each birth cohort, followed longitudinally, prevalence increases as children age (Fig. 2.5b): for children in the 2000 birth cohort, based on previous ASD prevalence estimates, at age 6 prevalence is 20/10,000, whereas at age 12, we may expect prevalence of 80/10,000 for the same birth cohort. Increasing prevalence rates with age within birth cohorts is unlikely to reflect the onset of ASD in later childhood and early adolescence. It is more likely that observed increases in prevalence reflect underdiagnosis in the preschool years as well as changes in public awareness, service availability, and diagnostic concepts and practices.

As an example, an analysis of special educational data from Minnesota showed a 16-fold increase in children identified with ASD from 1991–1992 to 2001–2002 (Gurney et al. 2003). However, during the same time period, an increase of 50% was observed for all disability categories (except severe intellectual deficiency); especially for the category including attention-deficit/hyperactivity disorder (ADHD). The large sample size allowed the authors to assess age, period, and cohort effects. Prevalence increased regularly in successive birth cohorts; for example, among 7-year-olds, prevalence rose from 18/10,000 among those born in 1989, to 29/10,000 among those born in 1991, to 55/10,000 in those born in 1993. Within the *same* birth cohorts, age effects were also apparent since for children born in 1989 the prevalence rose with age from 13/10,000 at age 6, to 21/10,000 at age 9, and 33/10,000 at age 11. As argued by Gurney et al. (2003), this pattern is not consistent with the natural etiology of ASD, which first manifests in early childhood. Gurney et al.'s analysis also showed a marked period effect, where rates started to increase in all ages and birth cohorts in the 1990s. The authors noted that this phenomenon coincided closely with the inclusion of ASDs in the federal Individuals with Disabilities Educational Act in the USA. A similar interpretation of upward trends had been put forward by Croen and colleagues (2002) in their analysis of the California department of developmental services data, and by Shattuck (2006) in his analysis of trends in US Department of Education data.

2.4 Conclusions

Epidemiological surveys of ASDs pose substantial challenges to researchers seeking to measure rates of ASD, particularly given the range of case definition, case identification, and case evaluation methods employed across surveys. However, from recent studies, a best estimate of (69/10,000) (equivalences = 6.9/1,000 or 0.69 % or 1 child in about 145 children) can be derived for the prevalence of ASD. Currently, the recent upward trend in rates of *prevalence* cannot be directly attributed to an increase in the *incidence* of the disorder, or to an epidemic of autism. Although power to detect time trends is seriously limited in existing datasets, there is good evidence that changes in diagnostic criteria and practices, policies for special education, service availability, and awareness of ASDs in both the lay and professional public may be responsible for increasing the prevalence over time. It is also noteworthy that the rise in number of children diagnosed occurred concurrently in many countries in the 1990s, when services for children with ASD also expanded significantly. Statistical power may also be a significant limitation in most investigations; thus, variations of small magnitude in ASD incidence may be undetected or should be interpreted with caution.

Nonetheless, the possibility that a true increase in the incidence of ASDs has also partially contributed to the upward trend in prevalence rates cannot, and should not, be completely eliminated based on the available data. To assess whether the incidence has increased, methodological factors that account for an important proportion of the variability in rates must be stringently controlled for. New survey methods have been developed for use in multinational comparisons; ongoing surveillance programs are currently underway and will soon provide more meaningful data to evaluate this hypothesis. Additionally, it remains to be seen how changes to diagnostic criteria introduced in the *DSM-5* will impact the ASD prevalence estimates going forward. Meanwhile, the available prevalence figures carry straightforward implications for current and future needs in services and early educational intervention programs.

References

Al-Farsi YM, Al-Sharbati MM, Al-Farsi OA, Al-Shafaee MS, Brooks DR, Waly MI (2011) Brief report: prevalence of autistic spectrum disorders in the Sultanate of Oman. J Autism Dev Disord 41(6):821–825. doi:10.1007/s10803-010-1094-8

American Psychiatric Association (1980) Diagnostic and statistical manual of mental disorders, 3rd ed. American Psychiatric Association, Washington, DC

American Psychiatric Association (1987) Diagnostic and statistical manual of mental disorders, (3rd ed, revision). American Psychiatric Association, Washington, DC

American Psychiatric Association (1994) Diagnostic and statistical manual of mental disorders, 4th ed. American Psychiatric Association, Washington, DC

American Psychiatric Association (2000). Diagnostic and statistical manual of mental disorders (4th ed, text revision). American Psychiatric Association, Washington, DC doi:10.1176/appi.books.9780890423349.

American Psychiatric Association (2013) Diagnostic and Statistical Manual of Mental Disorders, 5th ed. American Psychiatric Publishing., Washington, DC

Baird G, Charman T, Baron-Cohen S, Cox A, Swettenham J, Wheelwright S, Drew A (2000) A screening instrument for autism at 18 months of age: a 6-year follow-up study. J Am Acad Child Adolesc Psychiatry 39(6):694–702. doi:10.1097/00004583-200006000-00007

Baird G, Simonoff E, Pickles A, Chandler S, Loucas T, Meldrum D, Charman T (2006) Prevalence of disorders of the autism spectrum in a population cohort of children in South Thames: the special needs and autism project (SNAP). Lancet 368(9531):210–215. doi:10.1016/S0140-6736(06)69041-7

Brugha TS, McManus S, Bankart J, Scott F, Purdon S, Smith J et al (2011) Epidemiology of autism spectrum disorders in adults in the community in England. Arch Gen Psychiatry 68(5):459–465. doi:10.1001/archgenpsychiatry.2011.38

California Department of Developmental Services (2003) Autistic spectrum disorders: Changes in the California caseload-an update 1999 through 2002. Retrieved from http://www.dds.ca.gov/Autism/docs/AutismReport2003.pdf. Accessed April 2014

Centers for Disease Control and Prevention (2007a) Prevalence of autism spectrum disorders- autism and developmental disabilities monitoring network, six sites, United States, 2000. Morb Mortal Wkly Rep Surveillance Summaries, 56(1):1–11

Centers for Disease Control and Prevention (2007b) Prevalence of autism spectrum disorders – autism and developmental disabilities monitoring network, 14 Sites, United States, 2002. Morb Mortal Wkly Rep Surveillance Summaries, 56(SS01):12–28

Centers for Disease Control and Prevention (2009) Prevalence of autism spectrum disorders – autism and developmental disabilities monitoring network, United States, 2006. Morb Mortal Wkly Rep Surveillance Summaries, 58(SS10):1–20

Centers for Disease Control and Prevention (2012) Prevalence of autism spectrum disorders- autism and developmental disabilities monitoring network, 14 sites, United States, 2008. Morb Mortal Wkly Rep Surveillance Summaries, 61(3):1–19

Centers for Disease Control and Prevention (2014) Prevalence of autism spectrum disorders among children aged 8 years- autism and developmental disabilities monitoring network, 11 Sites, United States, 2010. Morb Mortal Wkly Rep Surveillance Summaries, 63(2):1–22

Chakrabarti S, Fombonne É (2001) Pervasive developmental disorders in preschool children. J Am Med Assoc 285(24):3093–3099. doi:10.1001/jama.285.24.3093

Chakrabarti S, Fombonne É (2005) Pervasive developmental disorders in preschool children: confirmation of high prevalence. Am J Psychiatry 162(6):1133–1141. doi:10.1176/appi.ajp.162.6.1133

Chien IC, Lin CH, Chou YJ, Chou P (2011) Prevalence and incidence of autism spectrum disorders among national health insurance enrollees in Taiwan from 1996 to 2005. J Child Neurol 26(7):830–834. doi:10.1177/0883073810393964

Croen LA, Grether JK, Hoogstrate J, Selvin S (2002) The changing prevalence of autism in California. J Autism Dev Disord 32(3):207–215
Croen LA, Najjar DV, Fireman B, Grether JK (2007) Maternal and paternal age and risk of autism spectrum disorders. Arch Pediatr Adolesc Med 161(4):334–340. doi:10.1001/archpedi.161.4.334
Davidovitch M, Hemo B, Manning-Courtney P, Fombonne É (2013) Prevalence and incidence of autism spectrum disorder in an Israeli population. J Autism Dev Disord 43(4):785–793. doi:10.1007/s10803-012-1611-z
Ellefsen A, Kampmann H, Billstedt E, Gillberg IC, Gillberg C (2007) Autism in the Faroe Islands: an epidemiological study. J Autism Dev Disord 37(3):437–444. doi:10.1007/s10803-006-0178-y
Elsabbagh M, Divan G, Koh Y-J, Kim YS, Kauchali S, Marcín C et al (2012) Global prevalence of autism and other pervasive developmental disorders. Autism Res 5(3):160–179. doi:10.1002/aur.239
Fombonne É (2001) Is there an epidemic of autism? Pediatrics 107(2):411–412
Fombonne É (2003a) Epidemiological surveys of autism and other pervasive developmental disorders: an update. J Autism Dev Disord 33(4):365–382
Fombonne É (2003b) The prevalence of autism. J Am Med Assoc 289(1):87–89. doi:10.1001/jama.289.1.87
Fombonne É (2005) Epidemiology of autistic disorder and other pervasive developmental disorders. J Clin Psychiatry 66(Suppl 10):3–8
Fombonne É (2007) Epidemiology. In Martin A, Volkmar F (eds) Lewis's child and adolescent psychiatry: a comprehensive textbook (4 ed). Philadelphia, Lippincott Williams & Wilkins, pp 150–171
Fombonne É (2009a) Epidemiology of pervasive developmental disorders. Pediatr Res 65(6):591–598. doi:10.1203/PDR.0b013e31819e7203
Fombonne É (2009b) Commentary: on King and Bearman. Int J Epidemiol 38(5):1241–1242. doi:10.1093/ije/dyp259
Fombonne É, Du Mazaubrun C (1992) Prevalence of infantile autism in four French regions. Soc Psychiatry Psychiatr Epidemiol 27(4):203–210. doi:10.1016/S0890-8567(09)66566-7
Fombonne É, Du Mazaubrun C, Cans C, Grandjean H (1997) Autism and associated medical disorders in a French epidemiological survey. J Am Acad Child Adolesc Psychiatry 36(11):1561–1569. doi:10.1016/S0890-8567(09)66566-7
Fombonne É, Zakarian R, Bennett A, Meng L, McLean-Heywood D (2006) Pervasive developmental disorders in Montreal, Quebec, Canada: prevalence and links with immunizations. Pediatrics 118(1):e139–e150. doi:10.1542/peds.2005-2993
Fombonne É, Quirke S, Hagen A (2011) Epidemiology of pervasive developmental disorders. In Amaral DG, Dawson G, Geschwind DH (eds) Autism spectrum disorders. New York, Oxford University Press, pp. 90–111
French L, Bertone A, Hyde K, Fombonne É (2013) Epidemiology of autism spectrum disorders. In Buxbaum JD, Hof PR (eds) The neuroscience of autism spectrum disorders. Oxford, Oxford University Press, pp 3–24
Gillberg C (1984) Infantile autism and other childhood psychoses in a Swedish urban region. Epidemiological aspects. J Child Psychol Psychiatry 25(1):35–43. doi:10.1111/j.1469-7610.1984.tb01717.x
Gillberg C, Steffenburg S, Schaumann H (1991) Is autism more common now than 10 years ago? Br J Psychiatry 158:403–409.
Gurney JG, Fritz MS, Ness KK, Sievers P, Newschaffer CJ, Shapiro EG (2003) Analysis of prevalence trends of autism spectrum disorder in Minnesota. Arch Pediatrics Adolesc Med 157(7):622–627. doi:10.1001/archpedi.157.7.622
Hill AP, Zuckerman KE, Fombonne É (2014) Epidemiology of Autism Spectrum Disorders. In Handbook of Autism and Pervasive Developmental Disorders. Hoboken, Wiley, pp 57–96
Honda H, Shimizu Y, Misumi K, Niimi M, Ohashi Y (1996) Cumulative incidence and prevalence of childhood autism in children in Japan. Br J Psychiatry 169(2):228–235

Honda H, Shimizu Y, Rutter M (2005) No effect of MMR withdrawal on the incidence of autism: a total population study. J Child Psychol Psychiatry 46(6):572–579. doi:10.1111/j.1469-7610.2005.01425.x

Idring S, Rai D, Dal H, Dalman C, Sturm H, Zander E et al (2012) Autism spectrum disorders in the Stockholm Youth Cohort: design, prevalence and validity. PLoS ONE 7(7):e41280. doi:10.1371/journal.pone.0041280

Kanner L (1943). Autistic disturbances of affective contact. Nerv child 2(3):217–250

Kawamura Y, Takahashi O, Ishii T (2008) Reevaluating the incidence of pervasive developmental disorders: impact of elevated rates of detection through implementation of an integrated system of screening in Toyota, Japan. Psychiatry Clin Neurosci 62(2):152–159. doi:10.1111/j.1440-1819.2008.01748.x

Kim YS, Leventhal BL, Koh Y-J, Fombonne É, Laska E, Lim E-C et al (2011) Prevalence of autism spectrum disorders in a total population sample. Am J Psychiatry 168(9):904–912. doi:10.1176/appi.ajp.2011.10101532

Kim YS, Fombonne É, Koh Y-J, Kim S-J, Cheon K-A, Leventhal BL (2014) A comparison of DSM-IV pervasive developmental disorder and DSM-5 autism spectrum disorder prevalence in an epidemiologic sample. J Am Acad Child Adolesc Psychiatry 53(5):500–508. doi:10.1016/j.jaac.2013.12.021

King M, Bearman P (2009) Diagnostic change and the increased prevalence of autism. Int J Epidemiol 38(5):1224–1234. doi:10.1093/ije/dyp261

Lauritsen MB, Pedersen CB, Mortensen PB (2004) The incidence and prevalence of pervasive developmental disorders: a Danish population-based study. Psychol Med 34(7):1339–1346.

Lazoff T, Zhong L, Piperni T, Fombonne É (2010) Prevalence of pervasive developmental disorders among children at the English montreal school board. Can J Psychiatry 55(11):715–720.

Leonard H, Glasson E, Nassar N, Whitehouse A, Bebbington A, Bourke J et al (2011) Autism and intellectual disability are differentially related to sociodemographic background at birth. PLoS ONE 6(3):e17875. doi:10.1371/journal.pone.0017875

Lord C, Rutter M, Couteur A (1994) Autism diagnostic interview-revised: a revised version of a diagnostic interview for caregivers of individuals with possible pervasive developmental disorders. J Autism Dev Disord 24(5):659–685. doi:10.1007/BF02172145

Lord C, Risi S, Lambrecht L, Cook EH Jr, Leventhal BL, DiLavore PC et al (2000) The autism diagnostic observation schedule-generic: a standard measure of social and communication deficits associated with the spectrum of autism. J Autism Dev Disord 30(3):205–223. doi:10.1023/A:1005592401947

Lotter V (1966) Epidemiology of autistic conditions in young children. Soc Psychiatry 1(3):124–135. doi:10.1007/bf00584048

Lotter V (1967) Epidemiology of autistic conditions in young children. Soc Psychiatry 1(4):163–173. doi:10.1007/BF00578950

Maenner MJ, Durkin MS (2010) Trends in the prevalence of autism on the basis of special education data. Pediatrics 126(5):e1018–e1025. doi:10.1542/peds.2010-1023

Maenner MJ, Rice CE, Arneson CL, Cunniff C, Schieve LA, Carpenter LA et al (2014) Potential impact of DSM-5 criteria on autism spectrum disorder prevalence estimates. JAMA Psychiatry 71(3):292–300. doi:10.1001/jamapsychiatry.2013.3893

Nassar N, Dixon G, Bourke J, Bower C, Glasson E, de Klerk N, Leonard H (2009) Autism spectrum disorders in young children: effect of changes in diagnostic practices. Int J Epidemiol 38(5):1245–1254. doi:10.1093/ije/dyp260

Nicholas JS, Charles JM, Carpenter LA, King LB, Jenner W, Spratt EG (2008) Prevalence and characteristics of children with autism-spectrum disorders. Ann Epidemiol 18(2):130–136. doi:10.1016/j.annepidem.2007.10.013

Nicholas JS, Carpenter LA, King LB, Jenner W, Charles JM (2009) Autism spectrum disorders in preschool-aged children: prevalence and comparison to a school-aged population. Ann Epidemiol 19(11):808–814. doi:10.1016/j.annepidem.2009.04.005

Nygren G, Cederlund M, Sandberg E, Gillstedt F, Arvidsson T, Carina Gillberg I et al (2012) The prevalence of autism spectrum disorders in toddlers: a population study of 2-year-old Swedish children. J Autism Dev Disord 42(7):1491–1497. doi:10.1007/s10803-011-1391-x

Ouellette-Kuntz H, Coo H, Yu CT, Chudley AE, Noonan A, Breitenbach M et al (2006) Prevalence of pervasive developmental disorders in two Canadian providences. J Policy Pract Intellect Disabil 3(3):164–172

Parner ET, Thorsen P, Dixon G, de Klerk N, Leonard H, Nassar N et al (2011) A comparison of autism prevalence trends in Denmark and Western Australia. J Autism Dev Disord 41(12):1601–1608. doi:10.1007/s10803-011-1186-0

Parner ET, Baron-Cohen S, Lauritsen MB, Jørgensen M, Schieve LA, Yeargin-Allsopp M, Obel C (2012) Parental age and autism spectrum disorders. Ann Epidemiol 22(3):143–150. doi:10.1016/j.annepidem.2011.12.006

Pinborough-Zimmerman J, Bakian AV, Fombonne É, Bilder D, Taylor J, McMahon WM (2012) Changes in the administrative prevalence of autism spectrum disorders: contribution of special education and health from 2002-2008. J Autism Dev Disord 42(4):521–530. doi:10.1007/s10803-011-1265-2

Posserud M, Lundervold AJ, Lie SA, Gillberg C (2010) The prevalence of autism spectrum disorders: impact of diagnostic instrument and non-response bias. Soc Psychiatry Psychiatr Epidemiol 45(3):319–327. doi:10.1007/s00127-009-0087-4

Rice CE, Baio J, Van Naarden Braun K, Doernberg N, Meaney FJ, Kirby RS, ADDM Network (2007) A public health collaboration for the surveillance of autism spectrum disorders. Paediatr Perinat Epidemiol 21(2):179–190. doi:10.1111/j.1365-3016.2007.00801.x

Rice C, Nicholas J, Baio J, Pettygrove S, Lee L-C, Van Naarden Braun K et al (2010) Changes in autism spectrum disorder prevalence in 4 areas of the United States. Disabil Health J 3(3):186–201. doi:10.1016/j.dhjo.2009.10.008

Rutter M (1970) Autistic children: infancy to adulthood. Semin Psychiatry 2(4):435–450

Saemundsen E, Magnusson P, Georgsdóttir I, Egilsson E, Rafnsson V (2013) Prevalence of autism spectrum disorders in an Icelandic birth cohort. BMJ Open 3(6). doi:10.1136/bmjopen-2013-002748

Samadi SA, Mahmoodizadeh A, McConkey R (2011) A national study of the prevalence of autism among five-year-old children in Iran. Autism 16(1):5–14. doi:10.1177/1362361311407091

Shattuck PT (2006) The contribution of diagnostic substitution to the growing administrative prevalence of autism in US special education. Pediatrics 117(4):1028–1037. doi:10.1542/peds.2005-1516

Taylor B, Miller E, Farrington CP, Petropoulos MC, Favot-Mayaud I, Li J, Waight PA (1999) Autism and measles, mumps, and rubella vaccine: no epidemiological evidence for a causal association. Lancet 353(9169):2026–2029.

Wazana A, Bresnahan M, Kline J (2007) The autism epidemic: fact or artifact? J Am Acad Child Adolesc Psychiatry 46(6):721–730

Williams JG, Higgins JPT, Brayne CEG (2006) Systematic review of prevalence studies of autism spectrum disorders. Arch Dis Child 91(1):8–15. doi:10.1136/adc.2004.062083

Windham GC, Anderson MC, Croen LA, Smith KS, Collins J, Grether JK (2011) Birth prevalence of autism spectrum disorders in the San Francisco Bay area by demographic and ascertainment source characteristics. J Autism Dev Disord 41(10):1362–1372. doi:10.1007/s10803-010-1160-2

World Health Organization (1977) International Statistical Classification of Diseases and Related Health Problems, Ninth Revision. World Health Organization, Geneva, Switzerland

World Health Organization (1992) International Statistical Classification of Diseases and Related Health Problems, Tenth Revision. World Health Organization, Geneva, Switzerland

Zahorodny W, Shenouda J, Howell S, Rosato NS, Peng B, Mehta U (2014) Increasing autism prevalence in metropolitan New Jersey. Autism 18(2):117–126. doi:10.1177/1362361312463977

Chapter 3
The Genetic Basis of Autism Spectrum Disorder

Jaqueline Bohrer Schuch, Luiza Monteavaro Mariath, Tatiana Roman and Lavinia Schuler-Faccini

Abstract Autism spectrum disorder (ASD) is multifactorial and complex condition, with a marked genetic influence, as evidenced by the high heritability (around 80–90%). Additionally, both common and rare genetic variants have an influence on the etiology and development. Several genetic syndromes are described in ASD, and they are present in approximately 10% of cases. Different genetic approaches are described in studies with ASD samples. Linkage studies mainly identify regions in chromosomes 2 and 7 which are involved in ASD. Genetic association studies, which aim to identify genes or genomic regions influencing ASD, involve the analysis of candidate genes or genome scan. The genes analyzed in these studies encode components related to neurotransmitter metabolism, neural migration, neuronal cell adhesion, apoptosis, and cell proliferation. Other methodologies have been used to complement genetic investigations, for example, exome and copy number variation (CNV) studies. Epigenetic studies indicate an important environmental influence on these disorders. The genetic evaluation of all patients with ASD is now required due to recent advances in knowledge regarding the genetic factors involved in the etiology of autism. Definitions of key terms and concepts used throughout this chapter are presented in Fig. 3.1.

Keywords Epigenetics · Gene · Autism

3.1 Introduction

Autism spectrum disorder (ASD) is a classic example of a complex multifactorial condition (Betancur 2011). Specific genetic etiology can be identified in only about 15% of ASD cases, which includes genetic factors and neurological or metabolic changes (Fig. 3.1).

Family studies suggest that the risk of recurrence among siblings is between 2 and 6%, which represents a 20–75 times greater prevalence than in the general population (Levy et al. 2009; Newschaffer et al. 2007). The high heritability (around

L. Schuler-Faccini (✉) · J. B. Schuch · L. M. Mariath · T. Roman
Genetics Department, Universidade Federal do Rio Grande do Sul, Porto Alegre, Brazil
e-mail: lavinia.faccini@ufrgs.br

M. d. l. A. Robinson-Agramonte (ed.), *Translational Approaches to Autism Spectrum Disorder*, DOI 10.1007/978-3-319-16321-5_3

Fig. 3.1 Definitions for key terms and concepts

Common variant/polymorphism: alteration in DNA sequence with higher frequency than 1% in the population.

Rare variant/mutation: alteration in DNA sequence with higher frequency less 1% in the population.

De novo mutation: not inherited mutation, mutation presents for the first time in one family member.

Penetrance: probability of an individual with given genetic variation manifest a characteristic associated with this variation.

Endophenotype: phenotypes "internal" or "intermediate" related to specific biological domains; or constructs, simpler than diseases or syndromes, possibly associated to genetic markers.

Epistasis: interaction between genes, where the presence of the gene may modify the effect of another gene.

80–90% in most studies) indicates a marked genetic influence in ASD etiology. Studies involving twins have shown concordance rates ranging from 60 to 92% among monozygotic twins and 0–10% among dizygotic twins (Ronald and Hoekstra 2011). However, Hallmayer et al. (2011) suggested that autism heritability is only moderate (around 38%) and that environmental factors have a significant influence on disease etiology. In this chapter, we will discuss the main genetic conditions and mechanisms involved in the etiology of autism.

3.1.1 Genetic Models

The study of the influence of different genetic factors (including common and rare variants) on multifactorial disorders like ASD follows different genetic models. The most commonly applied model is the *multifactorial polygenic model* that involves the contribution of several genes and environmental factors with small effects (Abrahams and Geschwind 2008). However, the *oligogenic model* could also be applied to explain the origin of ASD in some of the cases in which a few genes with moderate effects act together. The assessment of ASD domains (e.g., social, communication, and restricted and repetitive activities) shows only a modest covariation among them: individuals with extreme scores for the domains do not necessarily have similar scores in other areas. Nonetheless, these data suggest that the genetic susceptibility to ASD and to its phenotypic presentation results from the influence of several genes and their interactions, as well as the involvement of environmental factors (Ronald et al. 2006).

Family studies can also help the identification of unique genes with strong effects related to various aspects of the disease. Analysis of data from these studies indicated that some cases are associated with dominant de novo mutations (Abrahams and Geschwind 2008). These mutations appear to have a high penetrance in boys and low penetrance in girls, and also would act primarily in no familiar or sporadic ASD cases (Zhao et al. 2007). The identification of rare variants with a strong impact is extremely relevant to understanding ASD, and their phenotypic effect could be dependent on the genetic background in which rare variants occur (McClellan and King 2010).

The varied clinical manifestations present in ASD indicate that both common and rare genetic variants have an influence on its etiology and development. Based on current knowledge, the best genetic model for ASD should involve genetic analysis of more homogeneous subgroups sharing similar clinical characteristics. Additionally, the independent heritability in autism central areas is also important and should not be ignored (Anney et al. 2012).

3.2 Syndromes

Some genetic abnormalities are identified and associated with known genetic disorders, especially monogenic or chromosomal syndromes. However, until now, only around 10–15 % of ASD cases have been associated with known genetic factors. It is known that common biological bases, such as the disruption of molecular pathways and brain circuits, underlie different disorders, and this could explain the comorbidity of these syndromes with autism. Fragile X syndrome (FXS) is the most common, being identified in 3–5 % of ASD patients. Cytogenetic analyses may indicate the presence of known syndromes, transpositions, deletions, insertions, or other chromosomal rearrangements. Studies have estimated the frequency of these abnormalities in an additional 10 % of ASD sufferers (Zafeiriou et al. 2013).

3.2.1 Fragile X Syndrome (FXS)

FXS is considered to be the most common known condition of a single gene associated with ASD, affecting up to 5 % of cases. FXS is caused by expansion of the CGG trinucleotide repetition in the 5′-UTR region of the fragile X mental retardation (*FMR1*) gene (Wheeler et al. 2014; Steinberg and Webber 2013). The largest expansion of the CGG region—greater than 200 repetitions—is called a full mutation and it corresponds to FXS. Individuals with the *FMR1* pre-mutation have a CGG expansion ranging between 55 and 200 repetitions (McCary et al. 2013). FXS is characterized by intellectual disorder, anxiety, hyperactivity/impulsivity, and high rates of autism symptoms. Approximately 90 % of male individuals with FXS display at least one autistic behavioral feature (Brock and Hatton 2010). The prevalence of autism in FXS patients ranges from 15 to 52 % in different studies, depending on the diagnostic methods used (Zafeiriou et al. 2013).

The *FMR1* gene encodes the fragile X mental retardation protein (FMRP), which has an essential role in synaptic plasticity because it regulates the transport of mRNA and the translation in the brain, as well as regulates different pathways associated with autism (Liu and Takumi 2014). Previous studies have suggested that almost half of the genes involved in autism may be target genes of FMRP, leading to the hypothesis of the important role of FMRP in synaptic plasticity and in the development of autistic characteristics (Iossifov et al. 2012; Darnell et al. 2011). The absence of FMRP leads to significant changes in cognitive functions, dendritic spine morphology, and intracellular signaling. Studies have shown that children with FXS and ASD concomitantly exhibit weaker social and communicative skills, lower adaptive behavior, greater behavioral problems, and greater cognitive impairment than boys with FXS or idiopathic autism alone (Wheeler et al. 2014; Wolff et al. 2012).

3.2.2 Tuberous Sclerosis

Tuberous sclerosis complex (TSC) is an autosomal dominant disorder caused by inherited loss of function (30 %) or de novo (70 %) mutations in the *TSC1* or *TSC2* genes (Ghosh et al. 2013). Both these genes act as inhibitors in the target pathway of rapamycin (mTOR), a key pathway that controls the local translation at the synapse (Ghosh et al. 2013; Liu and Takumi 2014). The mTOR enzyme is an essential component of two complexes, mTORC1 and mTORC2, which play important roles in gene transcription, protein translation, and cell proliferation. Mutations in the *TSC1* or *TSC2* genes lead to excessive mTOR activity, which may cause disruption of the synaptic plasticity, characterized by excessive glutamate activity (Gipson et al. 2013).

TSC, one of the main causes of syndromic autism, is characterized by multisystemic hamartomas and debilitating neurological deficits, including seizures, intellectual disability, and ASD (Kothare et al. 2014). Autism has been reported in 26–61 % of individuals with TSC, with an average prevalence of 32 % (Gipson et al. 2013). Conversely, the prevalence of TSC among people with autism is 0–4 % (Fombonne 2003). People with TSC who develop ASD, display higher rates of intellectual disability, infantile spasms, and temporal lobe epilepsy (Gipson et al. 2013).

3.2.3 Rett Syndrome

Rett syndrome (RTT, OMIM 312750) is a neurodevelopmental disorder that occurs in about 1 in every 10,000 girls born. Up until the publication of DSM-IV, it had been classified as a pervasive developmental disorder. With the clear identification of the mutation causing RTT, it was excluded as an ASD in DSM-5.

RTT occurs most frequently in girls, although a small number of boys with the syndrome have been identified. Although approximately 95 % of RTT cases are caused by mutations in the methyl CpG-binding protein 2 (*MECP2*) gene, there is a well-established phenotypic heterogeneity. Approximately 80 % of the clinical cases of RTT present the typical clinical picture. After a seemingly normal development, at around 6–18 months of age, girls with RTT show neurological regression, lose their acquired cognitive, social, and motor skills, and develop autistic behavior accompanied by characteristic hand movements (Signorini et al. 2013). The autistic characteristics are typically temporary in RTT (Castro et al. 2013).

The nuclear protein MECP2 was initially associated with methylated DNA due to the binding of CpG islands (regions rich in guanines and cytosines). Additionally, a transcriptional repression domain that blocks the transcription of mRNA was identified in MECP2, which raised the initial hypothesis that MECP2 would only act as a regulating silencer of chromatin (Castro et al. 2013; Kishi and Macklis 2005). However, subsequent evidence has demonstrated a range of complex roles performed by MECP2. Besides also operating as a splicing modulator of RNA, MECP2 acts as a transcriptional activator by binding to the cAMP response element-binding (CREB) protein (Young et al. 2005; Chahrour et al. 2008) and playing a role in regulating the expression of microRNAs that are important for brain development and brain plasticity (Wu et al. 2010; Urdinguio et al. 2010). Studies have shown that through its different means of regulation, MECP2 is capable of indirectly controlling a range of molecules responsible for local protein synthesis and synaptic plasticity, which would be involved in RTT (Castro et al. 2013).

3.2.4 PTEN Hamartoma Tumor Syndrome (PHTS)

PTEN hamartoma tumor syndrome (PHTS) refers to a spectrum of disorders related to germline mutations in the homologous gene of phosphatase and tensin (*PTEN*), which include Cowden syndrome, Bannayan–Riley–Ruvalcaba syndrome, Lhermitte–Duclos disease, and the ASD associated with macrocephaly (Pilarski et al. 2013). The *PTEN* gene, which is located on chromosome 10, plays an important role in brain development. Studies indicate that germline mutations in this gene result in macrocephaly, seizures, and mental retardation (Lv et al. 2013). Subsequent biochemical studies have demonstrated that *PTEN* acts as a negative regulator of the phosphatidylinositol-3-kinase (PI3K)/AKT signaling pathway, which is an essential pathway for regulating the growth, survival, and proliferation of cells. *PTEN* acts by offsetting the action of PI3K, thereby inhibiting the activation of AKT and indirectly suppressing the mTOR pathway. Abnormalities in the PI3K/AKT/mTOR pathway lead to neurological and psychiatric disorders such as brain tumors, autism, and schizophrenia (Lv et al. 2013).

Studies on the frequency of *PTEN* mutations in patients with ASD indicate a variation of 1–17 %. It is possible that such discrepancies occur due to genetic variations among the populations of each study (Lv et al. 2013). Studies in mice have confirmed that deletions in *PTEN* result in long-term changes in social behavior, repetitive behavior, locomotor activity, and anxiety, resulting in many behavioral characteristics of autism (Lugo et al. 2014).

Besides the aforementioned syndromes, many others can also occur concomitantly with ASD. Table 3.1 presents a summary of the main syndromes associated with ASD.

3.3 Genetic Approaches

3.3.1 Linkage Studies

Linkage studies were a pioneering approach to the investigation of genetic factors in ASD. These studies evaluated families with individuals from different generations. The transmission of a chromosomal segment (locus) was verified from one generation to another, and cosegregation was sought between selected loci and the phenotype. This methodology allows identification of chromosomal regions, which may have more than one specific locus, thus suggesting the influence of different genes. The studies mainly identified regions in chromosomes 2 and 7, and also other regions in chromosomes 1, 5, 16, and 17 (Gupta and State 2006). More recently, other studies have indicated positive results in different regions in chromosomes 4, 6, 11, 15, 20, and 22 (Freitag et al. 2010; Fradin et al. 2010; Liu et al. 2008; Weiss et al. 2009). The 7q region is an area of particular interest, since many chromosomal rearrangements have been identified and some genes have been related to brain expression and other known physiological functions, possibly involved in the pathology of ASD (Gupta and State 2006).

3.3.2 Association Studies

Genetic association studies aim to identify genes or genomic regions that could influence a specific phenotype or disease. There are two main approaches that involve the study of: (1) candidate genes or (2) genome scans.

The candidate gene approach investigates common genetic variants located in genes associated with metabolic pathways and biological processes that are possibly related to phenotypes or diseases. There are two main methods to research: case control and family based. Case-control studies compare the allelic and genotypic frequencies in two groups of individuals: affected or unaffected by a given disease. Family-based studies observe the allele transmission from parents to affected offspring, comparing the data observed in the sample with the data expected by chance (Eley and Rijsdijk 2005).

Using these approaches, more than 100 genes have been evaluated, with both positive and negative results. Usually, the genes analyzed are those encoding

Table 3.1 Syndromes commonly associated to ASDs. (Source: Zafeiriou et al. 2013; Persico and Napolioni 2013; Abrahams and Geschwind 2008)

Syndrome	Gene/chromosomal region	Clinical characteristics	Prevalence in the population	Prevalence of the syndrome in the ASDs	Prevalence of ASD in the syndrome
Fragile X	*FMR1*	Intellectual disability, anxiety, hyperactivity/impulsiveness, autistic features	1/5160 boys	0–5 %	15–52 %
Tuberous sclerosis	*TSC1, TSC2*	Benign tumors in various organs; seizures, delayed development, behavioral problems	1/25,000	0–4 %	26–61 %
Rett	*MECP2*	Neurological regression, loss of acquired cognitive, social and motor abilities, autistic behaviors, stereotypical hand movements	1/10,000 girls	<1 %	100 %
PTEN Hamartoma tumor	*PTEN*	Macrocephaly, seizures intellectual disability	1/200,000–1/250,000	1–17 %	1–5 %
2q37 deletion	2q37.1–2q37.3	Mild to moderate intellectual disability, distinctive facial features, short stature, hypotonia, obesity, cardiac defects	Unknown	1.2 %	17–50 %
Angelman	*UBE3A* (15q11.2) maternal allele	Microcephaly, intellectual disability, severe impairment, movement and sleep disorders, characteristic behavior	1/40,000	<1 %	50–60 %
Prader–Willi	15q11–q13 paternal chromosome	Hypotonia, hyperphagia, obesity, short stature, motor and speech developmental delay, behavioral problems	1/25,000–1/45,000	<1 %	19–36, 5 %

Table 3.1 (continued)

Syndrome	Gene/chromosomal region	Clinical characteristics	Prevalence in the population	Prevalence of the syndrome in the ASDs	Prevalence of ASD in the syndrome
22q11.2 deletion DiGeorge	22q11.2 microdeletion	Congenital cardiac defect, cleft palate, immune deficiency, learning disabilities, distinctive facial features	1/1600–1/2000	<1%	14–50%
22q13 deletion Phelan-McDermid	22q13.3 deletion	Developmental delay, moderate to profound intellectual disability, hypotonia, and absent or delayed speech	Unknown	<1%	50–70%
Down	Chromosome 21 trisomy	Intellectual disabilities, motor developmental delay, characteristic facial features	1/1000	3.7%	15.6–19%
Neurofibromatosis type 1	*NF1*	Changes in skin pigmentation, benign tumors along nerves in the skin, brain, and other parts of the body	1/2500–1/3000	1.2%	4%
Turner	Chromosome X monosomy	Short stature, premature ovarian failure, webbed neck, lymphedema, coarctation of the aorta, normal intelligence	25–50/100,000 girls	Unknown	3%
Klinefelter	Extra X chromosome in males (47,XXY)	Small testes, low production of testosterone cryptorchidism, delayed or incomplete puberty, gynecomastia, infertility	40–172/100,000 boys	Unknown	11–27%
Smith–Lemli–Opitz	*DHCR7*	Craniofacial abnormalities, intellectual disabilities, behavior abnormalities	1/10,000–1/60,000	<1%	46–52%

Table 3.1 (continued)

Syndrome	Gene/chromosomal region	Clinical characteristics	Prevalence in the population	Prevalence of the syndrome in the ASDs	Prevalence of ASD in the syndrome
Cohen	*COH1*	Hypotonia, intellectual disabilities characteristic facial features, microcephaly	1/105,000	<1%	48%
Cornelia de Lange	*NIPBL, SMC1a and SMC3*	Slow growth before and after birth, intellectual disability that is usually severe to profound, skeletal abnormalities involving the arms and hands, and distinctive facial features	1/10,000	<1%	46–67%

components related to neurotransmitter metabolism (e.g., serotonin and glutamate), neural migration, neurodevelopmental control, neuronal cell adhesion, as well as genes involved in speech and language disorders, cognition, social interaction, and behavior. Also, chromosome break points in structural arrangements can be useful in identifying genes that have a major effect on ASD.

Some of the key candidate genes and common variants investigated in ASD studies are discussed below. They were chosen based on the AutismKB database (http://autismkb.cbi.pku.edu.cn/association_gene.php), considering the association score and the number of published studies.

3.3.2.1 Genes Involved in Neurotransmission

Serotonin is an essential neurotransmitter in brain functions such as sensory processing, cognition, imitation, communication, emotional and autonomic responses, regulation of motor activity, and it is also involved in the hormonal control of the hypothalamic and pituitary gland. The evaluation of serotonin in ASD is based on at least three major factors: its importance in neurodevelopment, its role in the treatment of symptoms, and the changes in blood/plasma levels in patients (Cook and Leventhal 1996; Whitaker-Azmitia 2005; Hollander et al. 2005). Neuroimaging studies have revealed that the peak capacity of brain serotonin synthesis occurs at around 2 years of age in children with typical development; however, this peak is not observed in children with autism (Chandana et al. 2005). The serotonin transporter (5HTT) is considered to be one of the main regulators of serotonergic function, and it is widely studied in ASD. The gene encoding the 5HTT (*SLC6A4*) is located on chromosome 17q11.1–q12. The 5HTTLPR (serotonin transporter-linked polymorphic region) is the most investigated variant among all described in the *SLC6A4* gene. The findings in several studies have been inconsistent—there were positive results for different alleles, and also negative data were observed. A more recent meta-analysis found no association between 5HTTLPR and ASD (Huang and Santangelo 2008). The assessment of clinical subgroups (endophenotypes) has been described in some studies, with different alleles being associated with deficits in communication and social interaction, as well as repetitive movements and aggressiveness (Brune et al. 2006).

The *GABRB3* gene is located in the 15q11–q13 region and it encodes one of the gamma-aminobutyric acid (GABA) receptors (Tierney et al. 2012), which is one of the largest inhibitory neurotransmitters in the mammalian nervous system. In this chromosomal region, there are two other genes that also encode GABA receptor subunits (*GABRA5* and *GABRG3*). According to linkage studies, such a region is considered to be a candidate for autism. Maternal or paternal deletions in the region are related to Angelman or Prader–Willi syndrome, respectively, which are diseases whose symptoms are associated with ASD (Buxbaum et al. 2002). The influence of the *GABRB3* gene in ASD has been evaluated in some studies, with some positive results already being detected (Buxbaum et al. 2002; McCauley et al. 2004; Curran et al. 2006).

The *SLC25A12* gene is located at 2q24, a region that is susceptible to ASD. It encodes a mitochondrial calcium-binding carrier protein, which is involved in the exchange of aspartate for glutamate via the inner mitochondrial membrane. It also regulates the cytosolic redox state and is an important supplier of energy to the neurons in the central nervous system (Durdiakova et al. 2014). Due to having an important role in pathways that are altered in autism, the *SLC25A12* gene has been indicated as one of the candidates for the disorder. Different common variants of the gene have been evaluated in several studies and some positive associations with autism have been shown (Silverman et al. 2008; Ramoz et al. 2004; Segurado et al. 2005).

The *GRIK2* gene encodes the glutamate 6 receptor and is considered to be a candidate because of its location at 6q21, a linkage region for autism. Glutamate receptors are the predominant excitatory neurotransmitter receptors in the brain and they are activated in a variety of normal neuropsychological processes. Due to their involvement in cognitive functions such as memory and learning, studies on the association of *GRIK2* with ASD were performed and some gene variants were positively associated (Shuang et al. 2004; Jamain et al. 2002; Kim et al. 2007).

3.3.2.2 Genes Involved in Cell Adhesion Molecules (CAMs)

Cell adhesion molecules (CAMs) mediate cell–cell and cell–extracellular matrix communication. They have an essential role in neurodevelopment, especially with regard to the formation and functioning of synapses, and they display specific expression patterns.

The β3 integrin (ITGB3) is the most investigated molecule in genetic studies of ASD, specifically involving synapses (Chavis and Westbrook 2001). Moreover, through different approaches, it was possible to infer that ITGB3 is associated with serotoninergic neurotransmission, due to its interaction with the serotonin transporter (Whyte et al. 2013; Ye et al. 2010). The *ITGB3* gene is located on chromosome 17q21.3. Different variants have been associated with ASD, either isolated or in gene–gene interactions, but mostly with the *SLC6A4* gene. The positive results involve different common variants across the studies (Weiss et al. 2006a; Schuch et al. 2014; Coutinho et al. 2007; Ma et al. 2010; Singh et al. 2013; Weiss et al. 2006b).

The gene from the neuronal CAM (*NRCAM*) is located at 7q31, a candidate region for autism, and it encodes a member of the CAM immunoglobulin superfamily. The NRCAM protein is present at high levels in the brain and in the spinal cord during development of the nervous system, and it interacts with other molecules to promote directional signaling during growth of the axonal cone (Marui et al. 2009). Some authors have investigated and found associations between variants in the *NRCAM* gene and ASD (Marui et al. 2009; Bonora et al. 2005).

3.3.2.3 Other Candidate Genes

The gene of the oxytocin receptor (*OXTR*) is located on the 3p25 chromosome and it encodes a protein from the family of receptors coupled to G-proteins. OXTR regulates the activity of the neuropeptide oxytocin (OXT), and both have been associated with the establishment of maternal and social behavior (Feldman et al. 2007; Ebstein et al. 2009; Di Napoli et al. 2014). Due to its role in social and emotional characteristics, the *OXTR* gene has been the target of several studies on the association with ASD, many of which have shown positive associations with different variants of the gene (LoParo and Waldman 2014).

The reelin (*RELN*) gene is located at chromosome 7q22 and it encodes a large protein from the extracellular matrix involved in the control of cell–cell interactions essential for cellular positioning and neuronal migration during brain development. The *RELN* gene is involved in different neurogenetic diseases such as autism, schizophrenia, bipolar disorder, and lissencephaly syndrome (Wang et al. 2014; Fatemi et al. 2000; Hong et al. 2000). Genetic variants have been positively associated with occurrences of autism in different studies (Serajee et al. 2006; Li et al. 2008; Ashley-Koch et al. 2007; Sharma et al. 2013).

The *MET* gene is located at 7q31 and it encodes the tyrosine kinase MET receptor, which is activated by the hepatocyte growth factor (HGF). Besides its initially established role as a proto-oncogene that operates in the development of various human cancers, *MET* also has a key role in early brain development, in the autoimmune system, and in gastrointestinal repair (Hedrick et al. 2012). The *MET* gene has been indicated as a candidate gene for autism due to its involvement in pathways related to the development of the brain cortex and cerebellum cortex (Zhou et al. 2011). Different common variants were evaluated in association studies, which showed SNP rs1858830 (C/G) in the *MET* promoter region as being one of the main gene variants associated with autism (Zhou et al. 2011).

The Engrailed-2 (*EN2*) gene is a homeobox transcription factor with a key role in the development of the midbrain and cerebellum. *EN2* was identified as a candidate gene for ASD, due to it being located at chromosome 7q36.3, a locus of susceptibility to ASD, and also because *EN2* mutant mice have exhibited morphological abnormalities of the cerebellum, similar to autistic individuals (Baader et al. 1998; Kuemerle et al. 1997; Wang et al. 2008).

However, most associations with candidate genes are not conclusive, and the data analysis and meta-analyses in various studies show conflicting and inconsistent findings. Although there are many studies showing positive associations, the majority was not replicated in different studies or populations.

The genome scan studies or genome-wide association studies (GWAS) evaluate, on a large scale, different genome variants, with no a priori definition of a biological hypothesis. This approach is very interesting for suggesting new genes and metabolic pathways in research on psychiatric disorders (like ASD), distinct from investigations of candidate genes. However, the common disease–common variant genetic approach that supports this kind of study may hinder the replication of results in many heterogeneous disorders. Few positive findings have been replicated independently across the studies—there is always a small effect associated with the risk of developing the disorder. The large number of markers analyzed in GWAS studies, which can surpass 500,000, increases the number of tests, resulting in the need for very strict data correction. The cutoff points used for statistical significance are low (e.g., $P < 5 \times 10^{-8}$) (Wang et al. 2009) and are very difficult to achieve in genes that have little effect.

Considering the latest published works, only the 5p14.1 region was associated with ASD in more than one investigation. This region encompasses the genes *CDH10* (cadherin 10, type 2) and *CDH9* (cadherin 9, type 2), which encode the cadherins responsible for cell–cell adhesion (Wang et al. 2009; Ma et al. 2009). Other GWAS performed so far only found suggestive or significant results that have not been replicated. The main findings of these studies involve genes related to CAMs, synaptogenesis, apoptosis, cell proliferation, coagulation, cancer, the neurotransmitter glutamate, and some intergenic regions (Anney et al. 2010; Jones et al. 2013; Ronald et al. 2010; Shi et al. 2013).

Common variants addressed by association studies generally have little effect on the disease etiology or characterization. The large complexity observed in heterogeneous populations and disorders is intimately connected to the size of the variable effect and the sample size investigated, sometimes hindering the study's findings. Moreover, the lack of functional evidence in some variants analyzed may also be contributing to this scenario. The use of other methodologies could, therefore, provide additional candidate genes for investigation.

3.3.3 Exome Studies

Exoma analysis involves the sequencing of all exons in the genome and it has been widely employed over the past few years. Its goal is to search for rare genetic variants associated with diseases. In ASD, several mutations have been described and associated with different biological mechanisms, such as chromatin structure, development of the nervous system, serotonin and glutamate neurotransmission, cell migration and organization, sodium channels, and ubiquitination (Bi et al. 2012; O'Roak et al. 2012; Neale et al. 2012; Toma et al. 2013; Sanders et al. 2012). Exoma analyses, like GWAS, have also been helpful in suggesting the involvement of new genes in the etiology of ASD. The results have also allowed, through use of bioinformatic tools, the description of complex biological pathways, integrating the new findings with the "classical" genes analyzed in studies with candidate genes, thus underscoring the importance of these studies.

3.3.4 Copy Number Variation (CNV) Studies

The evaluation of copy number variations (CNVs) involves the detection of rare variations in some regions on the genome. It has been widely used in psychiatric genetics. These variations may be duplications, or insertions or deletions (inherited or not), ranging in size from 1 kilobase (kb) to several megabases (Mb). Studies with different samples suggest that CNVs are associated with the normal variability of the human genome (Abecasis et al. 2012). However, they are considered to be a risk factor for some complex diseases, including ASD, and it is suggested that genes involved in behavioral aspects have a dose-dependent effect in accordance with the number of copies that are present.

In ASD, the CNVs are located throughout the genome at different chromosomes. The frequency of CNVs varies between 5 and 10% in ASD case subjects, and 1 and 2% in control subjects. Levy et al. (2011) analyzed the presence of CNVs in 28 ASD children and 62 healthy adults—they observed 38 CNVs. In all case subjects, at least one CNV was found, whereas in the control group the presence of such variants was found in only 11.3% of people. Moreover, 14 CNVs were found in both groups, 19 were observed only in ASD case subjects, and 5 were observed only in the control group. Even with the remarkable difference in the CNV frequencies between the case and control groups, it is important to note that there are common results in healthy and unhealthy individuals. Furthermore, similar results have recently been described in different psychiatric disorders—mostly in autism, schizophrenia, and cognitive deficits, including the de novo mutations (Stefansson et al. 2013).

Often, the duplication in the 15q11–13 region has been associated with ASD, mental deficiency, seizures, and behavioral and developmental abnormalities. Another alteration observed in approximately 1% of ASD cases is related to the 16p11.2 region, involving both duplications and deletions. Nonetheless, in both cases, these CNVs are not exclusive to ASD and can be observed in individuals diagnosed with other diseases. Also, the CNVs appear to not have complete penetrance and their pathogenic effect can vary according to the presence or absence of other genetic and environmental alterations (Stefansson et al. 2013). Some results have previously been associated with genetic syndromes, like Prader–Willi and DiGeorge (Glessner et al. 2009; Marshall et al. 2008; Pinto et al. 2010; Sanders et al. 2011; Sebat et al. 2007; Shishido et al. 2013). These studies are important to confirm or suggest new metabolic pathways such as neuronal cell adhesion and ubiquitin (Glessner et al. 2009).

The development of new techniques and methods in molecular and genetic research has resulted in an increase in information and knowledge regarding ASD, allowing new approaches to elucidate the etiology. It is known that the biological mechanisms involved in different phenotypes do not act in an isolated way, but interact with each other. The gene–gene interaction is a way of evaluating these epistatic effects. These interactions often involve prior biological hypotheses, constructed from biochemical and molecular studies. However, this approach has been little used in association studies (candidate gene or genome scan), and little solid evidence has been obtained. These more complex analyses may help us unravel the etiology of disorders like ASD, thus clarifying the biological processes involved.

Furthermore, an interesting study conducted by Poelmans et al. (2013) compiled the data obtained by studies using these different approaches in an attempt to make a more complete, comprehensive, and accurate interpretation of these bioinformatic analysis data. The authors evaluated the results of GWAS, CNV, exoma studies, and other genetic data. Through the enrichment of networks, important results involving the regulation of steroids, neurite outgrowth, and glutamatergic synapses were described. These analyses will be of great value, since the gathering and reviewing of results from different studies can highlight the most relevant information.

3.4 Epigenetics and Autism

Epigenetics refers to the processes that affect gene expression without altering the genetic code. Such epigenetic mechanisms may result in the silencing of specific genes and, therefore, they can affect the phenotype expressed (Mbadiwe and Millis 2013). Thus, epigenetics involves non-permanent hereditary modifications that alter gene expression without changing the DNA sequence itself. It is believed that the regulation of structure and neuronal function through epigenetic mechanisms is essential to the development of the central nervous system and that a change to this process may cause a range of neurodevelopmental disorders, including ASD. Epigenetic changes are partly a result of environmental effects, and, through modulation of the gene expression, environment plays an important role in the phenotype. The study of the role of epigenetics in autism has grown in recent years and today represents an important area of research (Rangasamy et al. 2013).

Some evidence points to a disruption of the global genome methylation in ASD, which would be an indication of the important influence that this modification has on the development of autism (Melnyk et al. 2012; Hall and Kelley 2014). The methylation mechanism is responsible for two other epigenetic processes involved in autism: the imprinting and the inactivation of the X chromosome. The imprinting concerns the epigenetic regulation of the gene expression in a manner dependent on parental origin. The imprinting controls, to a certain degree, if the genes expressed in the offspring will be from the maternal or paternal side (Hall and Kelley 2014). Studies have shown that many imprinted loci influence neuronal differentiation and the behavior or susceptibility to neurological diseases (Rangasamy et al. 2013). The inactivation of the X chromosome only occurs in females, when one of the copies of the X chromosome is deactivated via methylation to achieve the dosage compensation (Hall and Kelley 2014).

Environmental events that occur during early development can activate cellular signaling pathways associated with synaptic plasticity, even in stages after the development. In studies done on animals, the administration of valproic acid during the prenatal period or during weaning was associated with subsequent autistic behavior such as stereotyped behavior, lack of social interaction, increased anxiety, and cognitive and motor abnormalities. Valproic acid has known epigenetic effects (Tordjman et al. 2014; Vorhees 1987; Wu and Wang 2002).

Other environmental factors have also been studied in relation to the effects on DNA methylation. A study by Oh et al. (2013) showed that an adverse maternal environment induces methylation changes in the offspring's DNA, and that the affected genes encoded proteins are involved in the formation and functioning of synapses.

Epigenetic mechanisms have been reported in various genetic disorders associated with autism, including 15q11–q13 maternal duplication and various disorders such as FXS, Rett syndrome, Down syndrome, Turner syndrome, etc. Table 3.2 presents a summary of the main epigenetic mechanisms found in these syndromes associated with autism. The significant number of genetic disorders associated with epigenetic etiologies occurring concomitantly with autism indicates that epigenetic mechanisms involved in gene–environment interactions should be a common pathway for many ASD cases (Tordjman et al. 2014).

3.5 Conclusion

A genetic basis for autism is strongly supported by a large body of literature. Advances in clinical testing technology have increased the diagnostic yield from between 6 and 10% a few years ago to between 30 and 40% nowadays. Therefore, genetic testing should be discussed with all ASD patients and their families.

Guidelines for the genetic evaluation of patients with ASD were recently published (Schaefer et al. 2013).

Table 3.2 Epigenetic mechanisms observed in syndromes with ASDs. (Source: Rangasamy et al. 2013; Tordjman et al. 2014)

Genetic disorder	OMIM	Epigenetic mechanism
Fragile X syndrome (FXS)	#300624	DNA hypermethylation causing repression of *FMR1* gene; histone H3 and H4 tail deacetylation; histone H3–K9 methylation; and histone H3–K4 de-methylation
CHARGE syndrome	#214800	Chromatin remodeling
Maternal 15q11–q13 duplication	#608636	Possible disruptions of DNA methylation patterns and gene expression patterns within 15q11–q13 region
Angelman syndrome	#105830	Absent expression of maternally expressed gene *UBE3A* in brain due to lack of DNA methylation at maternal allele of imprinting center. The paternal allele of neuronal UBE3A is epigenetically silenced
Prader–Willi syndrome	#176270	Addition of DNA methylation at paternal allele of imprinting center suppresses expression of paternally expressed genes from imprinted cluster at 15q11–q13 region
Down syndrome	#190685	Trisomy for chromosome 21 results in over-expression of genes leading to abnormal brain development
Rett syndrome	#312750	De novo mutations or deletions disrupt binding to methyl-CpG, leading to abnormal gene transcription; disruptions in mRNA translation, histone methylation (H3K4, H3K9), and acetylation
Turner syndrome	–	Dosage changes in genes of the X chromosome that escape X chromosome inactivation. Potential imprinted gene on chromosome X

In summary, the literature indicates that the following approximate diagnostic tests are expected in the genetic evaluation of ASDs:

- CMA (10 %)
- Fragile X (1–5 %)
- MECP2 (4 % of females)
- PTEN (5 % of those with head circumferences >2.5 SDs that are tested)
- Karyotype (3 %)
- ACGH (30 %).

References

Abecasis GR, Auton A, Brooks LD, DePristo MA, Durbin RM, Handsaker RE, Kang HM, Marth GT, McVean GA (2012) An integrated map of genetic variation from 1092 human genomes. Nature 491(7422):56–65. doi:10.1038/nature11632

Abrahams BS, Geschwind DH (2008) Advances in autism genetics: on the threshold of a new neurobiology. Nature Rev Genet 9(5):341–355. doi:10.1038/nrg2346

Anney R, Klei L, Pinto D, Regan R, Conroy J, Magalhaes TR, Correia C, Abrahams BS, Sykes N, Pagnamenta AT, Almeida J, Bacchelli E, Bailey AJ, Baird G, Battaglia A, Berney T, Bolshakova N, Bolte S, Bolton PF, Bourgeron T, Brennan S, Brian J, Carson AR, Casallo G, Casey J, Chu SH, Cochrane L, Corsello C, Crawford EL, Crossett A, Dawson G, de Jonge M, Delorme R, Drmic I, Duketis E, Duque F, Estes A, Farrar P, Fernandez BA, Folstein SE, Fombonne E, Freitag CM, Gilbert J, Gillberg C, Glessner JT, Goldberg J, Green J, Guter SJ, Hakonarson H, Heron EA, Hill M, Holt R, Howe JL, Hughes G, Hus V, Igliozzi R, Kim C, Klauck SM, Kolevzon A, Korvatska O, Kustanovich V, Lajonchere CM, Lamb JA, Laskawiec M, Leboyer M, Le Couteur A, Leventhal BL, Lionel AC, Liu XQ, Lord C, Lotspeich L, Lund SC, Maestrini E, Mahoney W, Mantoulan C, Marshall CR, McConachie H, McDougle CJ, McGrath J, McMahon WM, Melhem NM, Merikangas A, Migita O, Minshew NJ, Mirza GK, Munson J, Nelson SF, Noakes C, Noor A, Nygren G, Oliveira G, Papanikolaou K, Parr JR, Parrini B, Paton T, Pickles A, Piven J, Posey DJ, Poustka A, Poustka F, Prasad A, Ragoussis J, Renshaw K, Rickaby J, Roberts W, Roeder K, Roge B, Rutter ML, Bierut LJ, Rice JP, Salt J, Sansom K, Sato D, Segurado R, Senman L, Shah N, Sheffield VC, Soorya L, Sousa I, Stoppioni V, Strawbridge C, Tancredi R, Tansey K, Thiruvahindrapduram B, Thompson AP, Thomson S, Tryfon A, Tsiantis J, Van Engeland H, Vincent JB, Volkmar F, Wallace S, Wang K, Wang Z, Wassink TH, Wing K, Wittemeyer K, Wood S, Yaspan BL, Zurawiecki D, Zwaigenbaum L, Betancur C, Buxbaum JD, Cantor RM, Cook EH, Coon H, Cuccaro ML, Gallagher L, Geschwind DH, Gill M, Haines JL, Miller J, Monaco AP, Nurnberger JI Jr, Paterson AD, Pericak-Vance MA, Schellenberg GD, Scherer SW, Sutcliffe JS, Szatmari P, Vicente AM, Vieland VJ, Wijsman EM, Devlin B, Ennis S, Hallmayer J (2010) A genome-wide scan for common alleles affecting risk for autism. Hum Mol Genet 19(20):4072–4082. doi:10.1093/hmg/ddq307

Anney R, Klei L, Pinto D, Almeida J, Bacchelli E, Baird G, Bolshakova N, Bolte S, Bolton PF, Bourgeron T, Brennan S, Brian J, Casey J, Conroy J, Correia C, Corsello C, Crawford EL, de Jonge M, Delorme R, Duketis E, Duque F, Estes A, Farrar P, Fernandez BA, Folstein SE, Fombonne E, Gilbert J, Gillberg C, Glessner JT, Green A, Green J, Guter SJ, Heron EA, Holt R, Howe JL, Hughes G, Hus V, Igliozzi R, Jacob S, Kenny GP, Kim C, Kolevzon A, Kustanovich V, Lajonchere CM, Lamb JA, Law-Smith M, Leboyer M, Le Couteur A, Leventhal BL, Liu XQ, Lombard F, Lord C, Lotspeich L, Lund SC, Magalhaes TR, Mantoulan C, McDougle CJ, Melhem NM, Merikangas A, Minshew NJ, Mirza GK, Munson J, Noakes C, Nygren G, Papanikolaou K, Pagnamenta AT, Parrini B, Paton T, Pickles A, Posey DJ, Poustka F, Ragoussis J, Regan R, Roberts W, Roeder K, Roge B, Rutter ML, Schlitt S, Shah N, Sheffield VC, Soorya L, Sousa I, Stoppioni V, Sykes N, Tancredi R, Thompson AP, Thomson S, Tryfon A, Tsiantis J, Van Engeland H, Vincent JB, Volkmar F, Vorstman JA, Wallace S, Wing K, Wittemeyer K, Wood S, Zurawiecki D, Zwaigenbaum L, Bailey AJ, Battaglia A, Cantor RM, Coon H, Cuccaro ML, Dawson G, Ennis S, Freitag CM, Geschwind DH, Haines JL, Klauck SM, McMahon WM, Maestrini E, Miller J, Monaco AP, Nelson SF, Nurnberger JI Jr, Oliveira G, Parr JR, Pericak-Vance MA, Piven J, Schellenberg GD, Scherer SW, Vicente AM, Wassink TH, Wijsman EM, Betancur C, Buxbaum JD, Cook EH, Gallagher L, Gill M, Hallmayer J, Paterson AD, Sutcliffe JS, Szatmari P, Vieland VJ, Hakonarson H, Devlin B (2012) Individual common variants exert weak effects on the risk for autism spectrum disorderspi. Human Mol Genet 21(21):4781–4792. doi:10.1093/hmg/dds301

Ashley-Koch AE, Jaworski J, Ma de Q, Mei H, Ritchie MD, Skaar DA, Robert Delong G, Worley G, Abramson RK, Wright HH, Cuccaro ML, Gilbert JR, Martin ER, Pericak-Vance MA (2007)

Investigation of potential gene-gene interactions between APOE and RELN contributing to autism risk. Psychiatr Genet 17(4):221–226. doi:10.1097/YPG.0b013e32809c2f75

Baader SL, Sanlioglu S, Berrebi AS, Parker-Thornburg J, Oberdick J (1998) Ectopic overexpression of engrailed-2 in cerebellar Purkinje cells causes restricted cell loss and retarded external germinal layer development at lobule junctions. J Neurosci 18(5):1763–1773

Betancur C (2011) Etiological heterogeneity in autism spectrum disorders: more than 100 genetic and genomic disorders and still counting. Brain Res 1380:42–77. doi:10.1016/j.brainres.2010.11.078

Bi C, Wu J, Jiang T, Liu Q, Cai W, Yu P, Cai T, Zhao M, Jiang YH, Sun ZS (2012) Mutations of ANK3 identified by exome sequencing are associated with autism susceptibility. Hum Mutat 33(12):1635–1638. doi:10.1002/humu.22174

Bonora E, Lamb JA, Barnby G, Sykes N, Moberly T, Beyer KS, Klauck SM, Poustka F, Bacchelli E, Blasi F, Maestrini E, Battaglia A, Haracopos D, Pedersen L, Isager T, Eriksen G, Viskum B, Sorensen EU, Brondum-Nielsen K, Cotterill R, Engeland H, Jonge M, Kemner C, Steggehuis K, Scherpenisse M, Rutter M, Bolton PF, Parr JR, Poustka A, Bailey AJ, Monaco AP (2005) Mutation screening and association analysis of six candidate genes for autism on chromosome 7q. Eur J Hum Genet 13(2):198–207. doi:10.1038/sj.ejhg.5201315

Brock M, Hatton D (2010) Distinguishing features of autism in boys with fragile X syndrome. J Intellect Disabil Res 54(10):894–905. doi:10.1111/j.1365–2788.2010.01315.x

Brune CW, Kim SJ, Salt J, Leventhal BL, Lord C, Cook EH Jr (2006) 5-HTTLPR Genotype-specific phenotype in children and adolescents with autism. Am J Psychiatry 163(12):2148–2156. doi:10.1176/appi.ajp.163.12.2148

Buxbaum JD, Silverman JM, Smith CJ, Greenberg DA, Kilifarski M, Reichert J, Cook EH Jr, Fang Y, Song CY, Vitale R (2002) Association between a GABRB3 polymorphism and autism. Mol Psychiatry 7(3):311–316. doi:10.1038/sj.mp.4001011

Castro J, Mellios N, Sur M (2013) Mechanisms and therapeutic challenges in autism spectrum disorders: insights from Rett syndrome. Curr Opin Neurol 26(2):154–159. doi:10.1097/WCO.0b013e32835f19a7

Chahrour M, Jung SY, Shaw C, Zhou X, Wong ST, Qin J, Zoghbi HY (2008) MeCP2, a key contributor to neurological disease, activates and represses transcription. Science 320(5880):1224–1229. doi:10.1126/science.1153252

Chandana SR, Behen ME, Juhasz C, Muzik O, Rothermel RD, Mangner TJ, Chakraborty PK, Chugani HT, Chugani DC (2005) Significance of abnormalities in developmental trajectory and asymmetry of cortical serotonin synthesis in autism. Int J Dev Neurosci 23(2–3):171–182. doi:10.1016/j.ijdevneu.2004.08.002

Chavis P, Westbrook G (2001) Integrins mediate functional pre- and postsynaptic maturation at a hippocampal synapse. Nature 411(6835):317–321. doi:10.1038/35077101

Cook EH, Leventhal BL (1996) The serotonin system in autism. Curr Opin Pediatr 8(4):348–354

Coutinho AM, Sousa I, Martins M, Correia C, Morgadinho T, Bento C, Marques C, Ataide A, Miguel TS, Moore JH, Oliveira G, Vicente AM (2007) Evidence for epistasis between SLC6A4 and ITGB3 in autism etiology and in the determination of platelet serotonin levels. Hum Genet 121(2):243–256. doi:10.1007/s00439-006-0301-3

Curran S, Powell J, Neale BM, Dworzynski K, Li T, Murphy D, Bolton PF (2006) An association analysis of candidate genes on chromosome 15 q11-13 and autism spectrum disorder. Mol Psychiatry 11(8):709–713. doi:10.1038/sj.mp.4001839 (England)

Darnell JC, Van Driesche SJ, Zhang C, Hung KY, Mele A, Fraser CE, Stone EF, Chen C, Fak JJ, Chi SW, Licatalosi DD, Richter JD, Darnell RB (2011) FMRP stalls ribosomal translocation on mRNAs linked to synaptic function and autism. Cell 146(2):247–261. doi:10.1016/j.cell.2011.06.013

Di Napoli A, Warrier V, Baron-Cohen S, Chakrabarti B (2014) Genetic variation in the oxytocin receptor (OXTR) gene is associated with Asperger syndrome. Mol Autism 5(1):48. doi:10.1186/2040-2392-5-48

Durdiakova J, Warrier V, Baron-Cohen S, Chakrabarti B (2014) Single nucleotide polymorphism rs6716901 in SLC25A12 gene is associated with Asperger syndrome. Mol Autism 5(1):25. doi:10.1186/2040-2392-5-25

Ebstein RP, Israel S, Lerer E, Uzefovsky F, Shalev I, Gritsenko I, Riebold M, Salomon S, Yirmiya N (2009) Arginine vasopressin and oxytocin modulate human social behavior. Ann N Y Acad Sci 1167:87–102. doi:10.1111/j.1749-6632.2009.04541.x

Eley TC, Rijsdijk F (2005) Introductory guide to the statistics of molecular genetics. J Child Psychol Psychiatry 46(10):1042–1044. doi:10.1111/j.1469-7610.2005.01523.x

Fatemi SH, Earle JA, McMenomy T (2000) Reduction in Reelin immunoreactivity in hippocampus of subjects with schizophrenia, bipolar disorder and major depression. Mol Psychiatry 5(6):654–663, 571

Feldman R, Weller A, Zagoory-Sharon O, Levine A (2007) Evidence for a neuroendocrinological foundation of human affiliation: plasma oxytocin levels across pregnancy and the postpartum period predict mother-infant bonding. Psychol Sci 18(11):965–970. doi:10.1111/j.1467-9280.2007.02010.x

Fombonne E (2003) Epidemiological surveys of autism and other pervasive developmental disorders: an update. J Autism Dev Disord 33(4):365–382

Fradin D, Cheslack-Postava K, Ladd-Acosta C, Newschaffer C, Chakravarti A, Arking DE, Feinberg A, Fallin MD (2010) Parent-of-origin effects in autism identified through genome-wide linkage analysis of 16,000 SNPs. PloS ONE 5(9). doi:10.1371/journal.pone.0012513

Freitag CM, Staal W, Klauck SM, Duketis E, Waltes R (2010) Genetics of autistic disorders: review and clinical implications. Eur Child Adolesc Psychiatry 19(3):169–178. doi:10.1007/s00787-009-0076-x

Ghosh A, Michalon A, Lindemann L, Fontoura P, Santarelli L (2013) Drug discovery for autism spectrum disorder: challenges and opportunities. Nat Rev Drug Discov 12(10):777–790. doi:10.1038/nrd4102

Gipson TT, Gerner G, Wilson MA, Blue ME, Johnston MV (2013) Potential for treatment of severe autism in tuberous sclerosis complex. World J Clin Pediatr 2(3):16–25. doi:10.5409/wjcp.v2.i3.16

Glessner JT, Wang K, Cai G, Korvatska O, Kim CE, Wood S, Zhang H, Estes A, Brune CW, Bradfield JP, Imielinski M, Frackelton EC, Reichert J, Crawford EL, Munson J, Sleiman PM, Chiavacci R, Annaiah K, Thomas K, Hou C, Glaberson W, Flory J, Otieno F, Garris M, Soorya L, Klei L, Piven J, Meyer KJ, Anagnostou E, Sakurai T, Game RM, Rudd DS, Zurawiecki D, McDougle CJ, Davis LK, Miller J, Posey DJ, Michaels S, Kolevzon A, Silverman JM, Bernier R, Levy SE, Schultz RT, Dawson G, Owley T, McMahon WM, Wassink TH, Sweeney JA, Nurnberger JI, Coon H, Sutcliffe JS, Minshew NJ, Grant SF, Bucan M, Cook EH, Buxbaum JD, Devlin B, Schellenberg GD, Hakonarson H (2009) Autism genome-wide copy number variation reveals ubiquitin and neuronal genes. Nature 459(7246):569–573. doi:10.1038/nature07953

Gupta AR, State MW (2006) [Autism: genetics]. Rev Bras Psiquiatr 28 (Suppl 1):S29–S38. doi:/S1516-44462006000500005

Hall L, Kelley E (2014) The contribution of epigenetics to understanding genetic factors in autism. Autism 18(8):872–881. doi:10.1177/1362361313503501

Hallmayer J, Cleveland S, Torres A, Phillips J, Cohen B, Torigoe T, Miller J, Fedele A, Collins J, Smith K, Lotspeich L, Croen LA, Ozonoff S, Lajonchere C, Grether JK, Risch N (2011) Genetic heritability and shared environmental factors among twin pairs with autism. Arch Gen Psychiatry 68(11):1095–1102. doi:10.1001/archgenpsychiatry.2011.76

Hedrick A, Lee Y, Wallace GL, Greenstein D, Clasen L, Giedd JN, Raznahan A (2012) Autism risk gene MET variation and cortical thickness in typically developing children and adolescents. Autism Res 5(6):434–439. doi:10.1002/aur.1256

Hollander E, Kaplan A, Schmeidler J, Yang H, Li D, Koran LM, Barbato LM (2005) Neurological soft signs as predictors of treatment response to selective serotonin reuptake inhibitors in obsessive-compulsive disorder. J Neuropsychiatry Clin Neurosci 17(4):472–477. doi:10.1176/appi.neuropsych.17.4.472

Hong SE, Shugart YY, Huang DT, Shahwan SA, Grant PE, Hourihane JO, Martin ND, Walsh CA (2000) Autosomal recessive lissencephaly with cerebellar hypoplasia is associated with human RELN mutations. Nat Genet 26(1):93–96. doi:10.1038/79246

Huang CH, Santangelo SL (2008) Autism and serotonin transporter gene polymorphisms: a systematic review and meta-analysis. Am J Med Genet B Neuropsychiatr Genet 147B(6):903–913. doi:10.1002/ajmg.b.30720

Iossifov I, Ronemus M, Levy D, Wang Z, Hakker I, Rosenbaum J, Yamrom B, Lee YH, Narzisi G, Leotta A, Kendall J, Grabowska E, Ma B, Marks S, Rodgers L, Stepansky A, Troge J, Andrews P, Bekritsky M, Pradhan K, Ghiban E, Kramer M, Parla J, Demeter R, Fulton LL, Fulton RS, Magrini VJ, Ye K, Darnell JC, Darnell RB, Mardis ER, Wilson RK, Schatz MC, McCombie WR, Wigler M (2012) De novo gene disruptions in children on the autistic spectrum. Neuron 74(2):285–299. doi:10.1016/j.neuron.2012.04.009

Jamain S, Betancur C, Quach H, Philippe A, Fellous M, Giros B, Gillberg C, Leboyer M, Bourgeron T (2002) Linkage and association of the glutamate receptor 6 gene with autism. Mol Psychiatry 7(3):302–310. doi:10.1038/sj.mp.4000979

Jones RM, Cadby G, Melton PE, Abraham LJ, Whitehouse AJ, Moses EK (2013) Genome-wide association study of autistic-like traits in a general population study of young adults. Front Hum Neurosci 7:658. doi:10.3389/fnhum.2013.00658

Kim SA, Kim JH, Park M, Cho IH, Yoo HJ (2007) Family-based association study between GRIK2 polymorphisms and autism spectrum disorders in the Korean trios. Neurosci Res 58(3):332–335. doi:10.1016/j.neures.2007.03.002

Kishi N, Macklis JD (2005) Dissecting MECP2 function in the central nervous system. J Child Neurol 20(9):753–759

Kothare SV, Singh K, Hochman T, Chalifoux JR, Staley BA, Weiner HL, Menzer K, Devinsky O (2014) Genotype/phenotype in tuberous sclerosis complex: associations with clinical and radiologic manifestations. Epilepsia 55(7):1020–1024. doi:10.1111/epi.12627

Kuemerle B, Zanjani H, Joyner A, Herrup K (1997) Pattern deformities and cell loss in Engrailed-2 mutant mice suggest two separate patterning events during cerebellar development. J Neurosci 17(20):7881–7889

Levy SE, Mandell DS, Schultz RT (2009) Autism. Lancet 374(9701):1627–1638. doi:10.1016/S0140-6736(09)61376-3

Levy D, Ronemus M, Yamrom B, Lee YH, Leotta A, Kendall J, Marks S, Lakshmi B, Pai D, Ye K, Buja A, Krieger A, Yoon S, Troge J, Rodgers L, Iossifov I, Wigler M (2011) Rare de novo and transmitted copy-number variation in autistic spectrum disorders. Neuron 70(5):886–897 doi:10.1016/j.neuron.2011.05.015

Li H, Li Y, Shao J, Li R, Qin Y, Xie C, Zhao Z (2008) The association analysis of RELN and GRM8 genes with autistic spectrum disorder in Chinese Han population. Am J Med Genet B Neuropsychiatr Genet 147b(2):194–200. doi:10.1002/ajmg.b.30584

Liu X, Takumi T (2014) Genomic and genetic aspects of autism spectrum disorder. Biochem Biophys Res Commun 452(2):244–253. doi:10.1016/j.bbrc.2014.08.108

Liu XQ, Paterson AD, Szatmari P (2008) Genome-wide linkage analyses of quantitative and categorical autism subphenotypes. Biol Psychiatry 64(7):561–570. doi:10.1016/j.biopsych.2008.05.023

LoParo D, Waldman ID (2014) The oxytocin receptor gene (OXTR) is associated with autism spectrum disorder: a meta-analysis. Mol Psychiatry. doi:10.1038/mp.2014.77

Lugo JN, Smith GD, Arbuckle EP, White J, Holley AJ, Floruta CM, Ahmed N, Gomez MC, Okonkwo O (2014) Deletion of PTEN produces autism-like behavioral deficits and alterations in synaptic proteins. Front Mol Neurosci 7:27. doi:10.3389/fnmol.2014.00027

Lv JW, Cheng TL, Qiu ZL, Zhou WH (2013) Role of the PTEN signaling pathway in autism spectrum disorder. Neurosci Bull 29(6):773–778. doi:10.1007/s12264-013-1382-3

Ma D, Salyakina D, Jaworski JM, Konidari I, Whitehead PL, Andersen AN, Hoffman JD, Slifer SH, Hedges DJ, Cukier HN, Griswold AJ, McCauley JL, Beecham GW, Wright HH, Abramson RK, Martin ER, Hussman JP, Gilbert JR, Cuccaro ML, Haines JL, Pericak-Vance MA (2009) A genome-wide association study of autism reveals a common novel risk locus at 5p14.1. Ann Hum Genet 73(Pt 3):263–273. doi:10.1111/j.1469-1809.2009.00523.x

Ma DQ, Rabionet R, Konidari I, Jaworski J, Cukier HN, Wright HH, Abramson RK, Gilbert JR, Cuccaro ML, Pericak-Vance MA, Martin ER (2010) Association and gene-gene interaction of

SLC6A4 and ITGB3 in autism. Am J Med Genet B Neuropsychiatr Genet 153B(2):477–483. doi:10.1002/ajmg.b.31003

Marshall CR, Noor A, Vincent JB, Lionel AC, Feuk L, Skaug J, Shago M, Moessner R, Pinto D, Ren Y, Thiruvahindrapduram B, Fiebig A, Schreiber S, Friedman J, Ketelaars CE, Vos YJ, Ficicioglu C, Kirkpatrick S, Nicolson R, Sloman L, Summers A, Gibbons CA, Teebi A, Chitayat D, Weksberg R, Thompson A, Vardy C, Crosbie V, Luscombe S, Baatjes R, Zwaigenbaum L, Roberts W, Fernandez B, Szatmari P, Scherer SW (2008) Structural variation of chromosomes in autism spectrum disorder. Am J Hum Genet 82(2):477–488. doi:10.1016/j.ajhg.2007.12.009

Marui T, Funatogawa I, Koishi S, Yamamoto K, Matsumoto H, Hashimoto O, Nanba E, Nishida H, Sugiyama T, Kasai K, Watanabe K, Kano Y, Sasaki T, Kato N (2009) Association of the neuronal cell adhesion molecule (NRCAM) gene variants with autism. Int J Neuropsychopharmacol 12(1):1–10. doi:10.1017/s1461145708009127

Mbadiwe T, Millis RM (2013) Epigenetics and autism. Autism Res Treat 2013:826156. doi:10.1155/2013/826156

McCary LM, Machlin L, Roberts JE (2013) The development of adaptive behavior in toddlers and preschoolers with fragile X versus autism. Int J Dev Disabil 59(2):67–79. doi:10.1179/2047387713y.0000000016

McCauley JL, Olson LM, Delahanty R, Amin T, Nurmi EL, Organ EL, Jacobs MM, Folstein SE, Haines JL, Sutcliffe JS (2004) A linkage disequilibrium map of the 1-Mb 15q12 GABA(A) receptor subunit cluster and association to autism. Am J Med Genet B Neuropsychiatr Genet 131b(1):51–59. doi:10.1002/ajmg.b.30038

McClellan J, King MC (2010) Genetic heterogeneity in human disease. Cell 141(2):210–217. doi:10.1016/j.cell.2010.03.032

Melnyk S, Fuchs GJ, Schulz E, Lopez M, Kahler SG, Fussell JJ, Bellando J, Pavliv O, Rose S, Seidel L, Gaylor DW, James SJ (2012) Metabolic imbalance associated with methylation dysregulation and oxidative damage in children with autism. J Autism Dev Disord 42(3):367–377. doi:10.1007/s10803-011-1260-7

Neale BM, Kou Y, Liu L, Ma'ayan A, Samocha KE, Sabo A, Lin CF, Stevens C, Wang LS, Makarov V, Polak P, Yoon S, Maguire J, Crawford EL, Campbell NG, Geller ET, Valladares O, Schafer C, Liu H, Zhao T, Cai G, Lihm J, Dannenfelser R, Jabado O, Peralta Z, Nagaswamy U, Muzny D, Reid JG, Newsham I, Wu Y, Lewis L, Han Y, Voight BF, Lim E, Rossin E, Kirby A, Flannick J, Fromer M, Shakir K, Fennell T, Garimella K, Banks E, Poplin R, Gabriel S, DePristo M, Wimbish JR, Boone BE, Levy SE, Betancur C, Sunyaev S, Boerwinkle E, Buxbaum JD, Cook EH Jr, Devlin B, Gibbs RA, Roeder K, Schellenberg GD, Sutcliffe JS, Daly MJ (2012) Patterns and rates of exonic de novo mutations in autism spectrum disorders. Nature 485(7397):242–245. doi:10.1038/nature11011

Newschaffer CJ, Croen LA, Daniels J, Giarelli E, Grether JK, Levy SE, Mandell DS, Miller LA, Pinto-Martin J, Reaven J, Reynolds AM, Rice CE, Schendel D, Windham GC (2007) The epidemiology of autism spectrum disorders. Annu Rev Public Health 28:235–258. doi:10.1146/annurev.publhealth.28.021406.144007

O'Roak BJ, Vives L, Girirajan S, Karakoc E, Krumm N, Coe BP, Levy R, Ko A, Lee C, Smith JD, Turner EH, Stanaway IB, Vernot B, Malig M, Baker C, Reilly B, Akey JM, Borenstein E, Rieder MJ, Nickerson DA, Bernier R, Shendure J, Eichler EE (2012) Sporadic autism exomes reveal a highly interconnected protein network of de novo mutations. Nature 485(7397):246–250. doi:10.1038/nature10989

Oh JE, Chambwe N, Klein S, Gal J, Andrews S, Gleason G, Shaknovich R, Melnick A, Campagne F, Toth M (2013) Differential gene body methylation and reduced expression of cell adhesion and neurotransmitter receptor genes in adverse maternal environment. Transl Psychiatry 3:e218. doi:10.1038/tp.2012.130

Persico AM, Napolioni V (2013) Autism genetics. Behavioural brain research 251:95–112. doi:10.1016/j.bbr.2013.06.012

Pilarski R, Burt R, Kohlman W, Pho L, Shannon KM, Swisher E (2013) Cowden syndrome and the PTEN hamartoma tumor syndrome: systematic review and revised diagnostic criteria. J Natl Cancer Inst 105(21):1607–1616. doi:10.1093/jnci/djt277

Pinto D, Pagnamenta AT, Klei L, Anney R, Merico D, Regan R, Conroy J, Magalhaes TR, Correia C, Abrahams BS, Almeida J, Bacchelli E, Bader GD, Bailey AJ, Baird G, Battaglia A, Berney T, Bolshakova N, Bolte S, Bolton PF, Bourgeron T, Brennan S, Brian J, Bryson SE, Carson AR, Casallo G, Casey J, Chung BH, Cochrane L, Corsello C, Crawford EL, Crossett A, Cytrynbaum C, Dawson G, de Jonge M, Delorme R, Drmic I, Duketis E, Duque F, Estes A, Farrar P, Fernandez BA, Folstein SE, Fombonne E, Freitag CM, Gilbert J, Gillberg C, Glessner JT, Goldberg J, Green A, Green J, Guter SJ, Hakonarson H, Heron EA, Hill M, Holt R, Howe JL, Hughes G, Hus V, Igliozzi R, Kim C, Klauck SM, Kolevzon A, Korvatska O, Kustanovich V, Lajonchere CM, Lamb JA, Laskawiec M, Leboyer M, Le Couteur A, Leventhal BL, Lionel AC, Liu XQ, Lord C, Lotspeich L, Lund SC, Maestrini E, Mahoney W, Mantoulan C, Marshall CR, McConachie H, McDougle CJ, McGrath J, McMahon WM, Merikangas A, Migita O, Minshew NJ, Mirza GK, Munson J, Nelson SF, Noakes C, Noor A, Nygren G, Oliveira G, Papanikolaou K, Parr JR, Parrini B, Paton T, Pickles A, Pilorge M, Piven J, Ponting CP, Posey DJ, Poustka A, Poustka F, Prasad A, Ragoussis J, Renshaw K, Rickaby J, Roberts W, Roeder K, Roge B, Rutter ML, Bierut LJ, Rice JP, Salt J, Sansom K, Sato D, Segurado R, Sequeira AF, Senman L, Shah N, Sheffield VC, Soorya L, Sousa I, Stein O, Sykes N, Stoppioni V, Strawbridge C, Tancredi R, Tansey K, Thiruvahindrapduram B, Thompson AP, Thomson S, Tryfon A, Tsiantis J, Van Engeland H, Vincent JB, Volkmar F, Wallace S, Wang K, Wang Z, Wassink TH, Webber C, Weksberg R, Wing K, Wittemeyer K, Wood S, Wu J, Yaspan BL, Zurawiecki D, Zwaigenbaum L, Buxbaum JD, Cantor RM, Cook EH, Coon H, Cuccaro ML, Devlin B, Ennis S, Gallagher L, Geschwind DH, Gill M, Haines JL, Hallmayer J, Miller J, Monaco AP, Nurnberger JI Jr, Paterson AD, Pericak-Vance MA, Schellenberg GD, Szatmari P, Vicente AM, Vieland VJ, Wijsman EM, Scherer SW, Sutcliffe JS, Betancur C (2010) Functional impact of global rare copy number variation in autism spectrum disorders. Nature 466(7304):368–372. doi:10.1038/nature09146

Poelmans G, Franke B, Pauls DD, Glennon JC, Buitelaar J (2013) AKAPs integrate genetic findings for autism spectrum disorders. Transl Psychiatry 3(6):e270. doi:10.1038/tp.2013.48

Ramoz N, Reichert JG, Smith CJ, Silverman JM, Bespalova IN, Davis KL, Buxbaum JD (2004) Linkage and association of the mitochondrial aspartate/glutamate carrier SLC25A12 gene with autism. Am J Psychiatry 161(4):662–669

Rangasamy S, D'Mello SR, Narayanan V (2013) Epigenetics, autism spectrum, and neurodevelopmental disorders. Neurotherapeutics 10(4):742–756. doi:10.1007/s13311-013-0227-0

Ronald A, Hoekstra RA (2011) Autism spectrum disorders and autistic traits: a decade of new twin studies. Am J Med Genet B Neuropsychiatr Genet 156(3):255–274

Ronald A, Happe F, Bolton P, Butcher LM, Price TS, Wheelwright S, Baron-Cohen S, Plomin R (2006) Genetic heterogeneity between the three components of the autism spectrum: a twin study. J Am Acad Child Adolesc Psychiatry 45(6):691–699. doi:10.1097/01.chi.0000215325.13058.9d

Ronald A, Butcher LM, Docherty S, Davis OS, Schalkwyk LC, Craig IW, Plomin R (2010) A genome-wide association study of social and non-social autistic-like traits in the general population using pooled DNA, 500 K SNP microarrays and both community and diagnosed autism replication samples. Behav Genet 40(1):31–45. doi:10.1007/s10519-009-9308-6

Sanders SJ, Ercan-Sencicek AG, Hus V, Luo R, Murtha MT, Moreno-De-Luca D, Chu SH, Moreau MP, Gupta AR, Thomson SA, Mason CE, Bilguvar K, Celestino-Soper PB, Choi M, Crawford EL, Davis L, Wright NR, Dhodapkar RM, DiCola M, DiLullo NM, Fernandez TV, Fielding-Singh V, Fishman DO, Frahm S, Garagaloyan R, Goh GS, Kammela S, Klei L, Lowe JK, Lund SC, McGrew AD, Meyer KA, Moffat WJ, Murdoch JD, O'Roak BJ, Ober GT, Pottenger RS, Raubeson MJ, Song Y, Wang Q, Yaspan BL, Yu TW, Yurkiewicz IR, Beaudet AL, Cantor RM, Curland M, Grice DE, Gunel M, Lifton RP, Mane SM, Martin DM, Shaw CA, Sheldon M, Tischfield JA, Walsh CA, Morrow EM, Ledbetter DH, Fombonne E, Lord C, Martin CL, Brooks AI, Sutcliffe JS, Cook EH Jr, Geschwind D, Roeder K, Devlin B, State MW (2011) Multiple recurrent de novo CNVs, including duplications of the 7q11.23 Williams syndrome region, are strongly associated with autism. Neuron 70(5):863–885. doi:10.1016/j.neuron.2011.05.002

Sanders SJ, Murtha MT, Gupta AR, Murdoch JD, Raubeson MJ, Willsey AJ, Ercan-Sencicek AG, DiLullo NM, Parikshak NN, Stein JL, Walker MF, Ober GT, Teran NA, Song Y, El-Fishawy P, Murtha RC, Choi M, Overton JD, Bjornson RD, Carriero NJ, Meyer KA, Bilguvar K, Mane SM, Sestan N, Lifton RP, Gunel M, Roeder K, Geschwind DH, Devlin B, State MW (2012) De novo mutations revealed by whole-exome sequencing are strongly associated with autism. Nature 485(7397):237–241. doi:10.1038/nature10945

Schaefer GB, Mendelsohn NJ, Professional Pratice and Guidelines Committee (2013) Clinical genetics evaluation in identifying the etiology of autism spectrum disorders: 2013 guideline revisions. Genet Med 15(5):399–407. doi:10.1038/gim.2013.32

Schuch JB, Muller D, Endres RG, Bosa CA, Longo D, Schuler-Faccini L, Ranzan J, Becker MM, dos Santos Riesgo R, Roman T (2014) The role of b3 integrin gene variants in autism spectrum disorders—diagnosis and symptomatology. Gene 553:24–30

Sebat J, Lakshmi B, Malhotra D, Troge J, Lese-Martin C, Walsh T, Yamrom B, Yoon S, Krasnitz A, Kendall J, Leotta A, Pai D, Zhang R, Lee YH, Hicks J, Spence SJ, Lee AT, Puura K, Lehtimaki T, Ledbetter D, Gregersen PK, Bregman J, Sutcliffe JS, Jobanputra V, Chung W, Warburton D, King MC, Skuse D, Geschwind DH, Gilliam TC, Ye K, Wigler M (2007) Strong association of de novo copy number mutations with autism. Science 316(5823):445–449. doi:10.1126/science.1138659

Segurado R, Conroy J, Meally E, Fitzgerald M, Gill M, Gallagher L (2005) Confirmation of association between autism and the mitochondrial aspartate/glutamate carrier SLC25A12 gene on chromosome 2q31. Am J Psychiatry 162(11):2182–2184. doi:10.1176/appi.ajp.162.11.2182

Serajee FJ, Zhong H, Mahbubul Huq AH (2006) Association of Reelin gene polymorphisms with autism. Genomics 87(1):75–83. doi:10.1016/j.ygeno.2005.09.008

Sharma JR, Arieff Z, Gameeldien H, Davids M, Kaur M, van der Merwe L (2013) Association analysis of two single-nucleotide polymorphisms of the RELN gene with autism in the South African population. Genet Test Mol Biomarkers 17(2):93–98. doi:10.1089/gtmb.2012.0212

Shi L, Zhang X, Golhar R, Otieno FG, He M, Hou C, Kim C, Keating B, Lyon GJ, Wang K, Hakonarson H (2013) Whole-genome sequencing in an autism multiplex family. Mol Autism 4(1):8. doi:10.1186/2040-2392-4-8

Shishido E, Aleksic B, Ozaki N (2013) Copy-number variation in the pathogenesis of autism spectrum disorder. Psychiatry Clin Neurosci. doi:10.1111/pcn.12128

Shuang M, Liu J, Jia MX, Yang JZ, Wu SP, Gong XH, Ling YS, Ruan Y, Yang XL, Zhang D (2004) Family-based association study between autism and glutamate receptor 6 gene in Chinese Han trios. Am J Med Genet B Neuropsychiatr Genet 131b(1):48–50. doi:10.1002/ajmg.b.30025

Signorini C, De Felice C, Durand T, Oger C, Galano JM, Leoncini S, Pecorelli A, Valacchi G, Ciccoli L, Hayek J (2013) Isoprostanes and 4-hydroxy-2-nonenal: markers or mediators of disease? Focus on Rett syndrome as a model of autism spectrum disorder. Oxid Med Cell Longev 2013:343824. doi:10.1155/2013/343824

Silverman JM, Buxbaum JD, Ramoz N, Schmeidler J, Reichenberg A, Hollander E, Angelo G, Smith CJ, Kryzak LA (2008) Autism-related routines and rituals associated with a mitochondrial aspartate/glutamate carrier SLC25A12 polymorphism. Am J Med Genet B Neuropsychiatr Genet 147(3):408–410. doi:10.1002/ajmg.b.30614

Singh AS, Chandra R, Guhathakurta S, Sinha S, Chatterjee A, Ahmed S, Ghosh S, Rajamma U (2013) Genetic association and gene-gene interaction analyses suggest likely involvement of ITGB3 and TPH2 with autism spectrum disorder (ASD) in the Indian population. Prog Neuropsychopharmacol Biol Psychiatry 45:131–143. doi:10.1016/j.pnpbp.2013.04.015

Stefansson H, Meyer-Lindenberg A, Steinberg S, Magnusdottir B, Morgen K, Arnarsdottir S, Bjornsdottir G, Walters GB, Jonsdottir G, Doyle OM, Tost H, Grimm O, Kristjansdottir S, Snorrason H, Davidsdottir SR, Gudmundsson LJ, Jonsson GF, Stefansdottir B, Helgadottir I, Haraldsson M, Jonsdottir B, Thygesen JH, Schwarz AJ, Didriksen M, Stensbol TB, Brammer M, Kapur S, Halldorsson JG, Hreidarsson S, Saemundsen E, Sigurdsson E, Stefansson K (2013) CNVs conferring risk of autism or schizophrenia affect cognition in controls. Nature. doi:10.1038/nature12818

Steinberg J, Webber C (2013) The roles of FMRP-regulated genes in autism spectrum disorder: single- and multiple-hit genetic etiologies. Am J Hum Genet 93(5):825–839. doi:10.1016/j.ajhg.2013.09.013

Tierney AL, Gabard-Durnam L, Vogel-Farley V, Tager-Flusberg H, Nelson CA (2012) Developmental trajectories of resting EEG power: an endophenotype of autism spectrum disorder. PloS ONE 7(6):e39127. doi:10.1371/journal.pone.0039127

Toma C, Torrico B, Hervas A, Valdes-Mas R, Tristan-Noguero A, Padillo V, Maristany M, Salgado M, Arenas C, Puente XS, Bayes M, Cormand B (2013) Exome sequencing in multiplex autism families suggests a major role for heterozygous truncating mutations. Mol Psychiatry. doi:10.1038/mp.2013.106

Tordjman S, Somogyi E, Coulon N, Kermarrec S, Cohen D, Bronsard G, Bonnot O, Weismann-Arcache C, Botbol M, Lauth B, Ginchat V, Roubertoux P, Barburoth M, Kovess V, Geoffray MM, Xavier J (2014) Gene x environment interactions in autism spectrum disorders: role of epigenetic mechanisms. Front Psychiatry 5:53. doi:10.3389/fpsyt.2014.00053

Urdinguio RG, Fernandez AF, Lopez-Nieva P, Rossi S, Huertas D, Kulis M, Liu CG, Croce CM, Calin GA, Esteller M (2010) Disrupted microRNA expression caused by Mecp2 loss in a mouse model of Rett syndrome. Epigenetics 5(7):656–663. doi:10.4161/epi.5.7.13055

Vorhees CV (1987) Behavioral teratogenicity of valproic acid: selective effects on behavior after prenatal exposure to rats. Psychopharmacology (Berl) 92(2):173–179

Wang L, Jia M, Yue W, Tang F, Qu M, Ruan Y, Lu T, Zhang H, Yan H, Liu J, Guo Y, Zhang J, Yang X, Zhang D (2008) Association of the ENGRAILED 2 (EN2) gene with autism in Chinese Han population. Am J Med Genet B Neuropsychiatr Genet 147b(4):434–438. doi:10.1002/ajmg.b.30623

Wang K, Zhang H, Ma D, Bucan M, Glessner JT, Abrahams BS, Salyakina D, Imielinski M, Bradfield JP, Sleiman PM (2009) Common genetic variants on 5p14. 1 associate with autism spectrum disorders. Nature 459(7246):528–533

Wang Z, Hong Y, Zou L, Zhong R, Zhu B, Shen N, Chen W, Lou J, Ke J, Zhang T, Wang W, Miao X (2014) Reelin gene variants and risk of autism spectrum disorders: an integrated meta-analysis. Am J Med Genet B Neuropsychiatr Genet 165b(2):192–200. doi:10.1002/ajmg.b.32222

Weiss LA, Kosova G, Delahanty RJ, Jiang L, Cook EH, Ober C, Sutcliffe JS (2006a) Variation in ITGB3 is associated with whole-blood serotonin level and autism susceptibility. Eur J Hum Genet 14(8):923–931. doi:10.1038/sj.ejhg.5201644

Weiss LA, Ober C, Cook EH Jr (2006b) ITGB3 shows genetic and expression interaction with SLC6A4. Hum Genet 120(1):93–100. doi:10.1007/s00439-006-0196-z

Weiss LA, Arking DE, Daly MJ, Chakravarti A (2009) A genome-wide linkage and association scan reveals novel loci for autism. Nature 461(7265):802–808. doi:10.1038/nature08490

Wheeler AC, Mussey J, Villagomez A, Bishop E, Raspa M, Edwards A, Bodfish J, Bann C, Bailey DB Jr (2014) DSM-5 changes and the prevalence of parent-reported autism spectrum symptoms in fragile X syndrome. J Autism Dev Disord. doi:10.1007/s10803-014-2246-z

Whitaker-Azmitia PM (2005) Behavioral and cellular consequences of increasing serotonergic activity during brain development: a role in autism? Int J Dev Neurosci 23(1):75–83. doi:10.1016/j.ijdevneu.2004.07.022

Whyte A, Jessen T, Varney S, Carneiro AM (2013) Serotonin transporter and integrin beta 3 genes interact to modulate serotonin uptake in mouse brain. Neurochem Int. doi:10.1016/j.neuint.2013.09.014

Wolff JJ, Bodfish JW, Hazlett HC, Lightbody AA, Reiss AL, Piven J (2012) Evidence of a distinct behavioral phenotype in young boys with fragile X syndrome and autism. J Am Acad Child Adolesc Psychiatry 51(12):1324–1332. doi:10.1016/j.jaac.2012.09.001

Wu Y, Wang L (2002) The effects of antiepileptic drugs on spatial learning and hippocampal protein kinase C gamma in immature rats. Brain Dev 24(2):82–87

Wu H, Tao J, Chen PJ, Shahab A, Ge W, Hart RP, Ruan X, Ruan Y, Sun YE (2010) Genome-wide analysis reveals methyl-CpG-binding protein 2-dependent regulation of microRNAs in a mouse model of Rett syndrome. Proc Natl Acad Sci U S A 107(42):18161–18166. doi:10.1073/pnas.1005595107

Ye H, Liu J, Wu JY (2010) Cell adhesion molecules and their involvement in autism spectrum disorder. Neurosignals 18(2):62–71. doi:10.1159/000322543

Young JI, Hong EP, Castle JC, Crespo-Barreto J, Bowman AB, Rose MF, Kang D, Richman R, Johnson JM, Berget S, Zoghbi HY (2005) Regulation of RNA splicing by the methylation-dependent transcriptional repressor methyl-CpG binding protein 2. Proc Natl Acad Sci U S A 102(49):17551–17558. doi:10.1073/pnas.0507856102

Zafeiriou DI, Ververi A, Dafoulis V, Kalyva E, Vargiami E (2013) Autism spectrum disorders: the quest for genetic syndromes. Am J Med Genet B Neuropsychiatr Genet 162b(4):327–366. doi:10.1002/ajmg.b.32152

Zhao X, Leotta A, Kustanovich V, Lajonchere C, Geschwind DH, Law K, Law P, Qiu S, Lord C, Sebat J, Ye K, Wigler M (2007) A unified genetic theory for sporadic and inherited autism. Proc Natl Acad Sci U S A 104(31):12831–12836. doi:10.1073/pnas.0705803104

Zhou X, Xu Y, Wang J, Zhou H, Liu X, Ayub Q, Wang X, Tyler-Smith C, Wu L, Xue Y (2011) Replication of the association of a MET variant with autism in a Chinese Han population. PLoS ONE 6(11):e27428. doi:10.1371/journal.pone.0027428

Chapter 4
Autism Spectrum Disorder. A Clinical Neurophysiology Approach I

Lilia María Morales Chacón and Margarita Minou Báez Martin

Abstract The lack of well-controlled studies on treatment efficacy, along with the immense costs of education and care for patients with autism spectrum disorder (ASD) makes ASD a public health problem. Likewise, the enormous heterogeneity of the clinical and behavioral symptoms has made it rather difficult to delineate the neural circuitry affiliated with this condition. The objective of this review is to provide a summary of recent data on the clinical neurophysiology in patients with ASD. Electroencephalographic (EEG) findings with emphasis on quantitative EEG (power EEG, coherence EEG) are summarized since EEG studies provide evidence of brain functional aspects in this disorder. Besides addressing current knowledge of the relationship among EEG abnormalities, epilepsy, and regression in ASD, this chapter further highlights the most recent literature on sleep disorders and polysomnography (PSG) findings in this pathology. Lastly, we discuss limitations in available research that may contribute to understand the inconsistencies in the literature, and offer suggestions for future research in this area for advancing the understanding of ASD.

Keywords Autism spectrum disorder · Electroencephalography · Quantitative electroencephalography · Epileptiform abnormalities · Epilepsy · Autistic regression · Polysomnography

Abbreviations

ASD	Autism spectrum disorder
EA	Epileptiform abnormalities
EEG	Electroencephalography
MEG	Magnetoencephalography
MNS	Mirror neurons system
MRI	Magnetic resonance imaging

L. M. Morales Chacón (✉)
International Center for Neurological Restoration (CIREN), Clinical Neurophysiology Service, Ave 25 15805 e/158 y 160, Playa, 11300 Havana, Cuba
e-mail: lilia.morales@infomed.sld.cu

M. M. Báez Martin
e-mail: minou@neuro.ciren.cu

M. d. l. A. Robinson-Agramonte (ed.), *Translational Approaches to Autism Spectrum Disorder*, DOI 10.1007/978-3-319-16321-5_4

NF	Neurofeedback
PSG	Polysomnography
QEEG	Quantitative electroencephalography

4.1 Introduction

Autism spectrum disorder (ASD) is a new *Diagnostic and Statistical Manual of Mental Disorders, Fifth Edition (DSM-5)* disorder encompassing the previous *Diagnostic and Statistical Manual of Mental Disorders, Fourth Edition (DSMIV)* autistic disorder (autism), Asperger's disorder, childhood disintegrative disorder, Rett's disorder, and pervasive developmental disorder not otherwise specified. It is characterized by deficits in two core domains: deficits in social communication and social interaction and restricted repetitive patterns of behavior, interests, and activities (Volkmar and McPartland 2014; Kelly 2014). This pathology is a behaviorally defined syndrome of early life, with varied symptoms reflecting many biological and environmental influences that are unique to each individual's brain and that shape its unique developmental trajectory (Jeste 2011; Toth and Stobbe 2011).

ASD has been associated with disturbance in cerebral organization and function (Palau-Baduell et al. 2013). The goal of neurologic, genetic, electrophysiologic, imaging, and other biologic tests performed in this disease is not only to diagnose autism but to attempt to define some of its many potential etiologies and understand their pathogenic effects on the brain (Silver and Rapin 2012).

The objective of this review is to present research data on the electroencephalography (EEG) findings, with emphasis on quantitative EEG (power EEG, coherence EEG), epileptiform EEG abnormalities and sleep disorders in patients with ASD. Besides addressing current knowledge of the relationship among EEG abnormalities, epilepsy, and regression in ASD this chapter further highlights the most recent literature on sleep disorders and PSG findings in this pathology.

4.2 Electroencephalography (EEG) and Autism Spectrum Disorder (ASD)

ASD involving multiple neural system dysfunctions and EEG studies provide evidence of brain functional aspects in this disorder (Palau-Baduell et al. 2011; Palau-Baduell et al. 2013).

The first EEG study in autism reported EEG abnormalities in 58 % of the autistic children studied (White et al. 1964); later research focused on epileptiform activity. However, other EEG findings have been reported such as: suppression of mu waves, reduced alpha power, photic driving on the right, and low levels of coherence and synchronization. Some of these EEG results are discussed below.

Mu rhythm at 7–12 Hz can be recorded from the central scalp areas and is reactive in the normal population (mu suppression) by movements or thoughts of movement and also when the subject is observing others' movements. The latter refers to the mirror neurons system (MNS) and is normally active when observing others, proposed as deficient in autism (Hughes 2010). Mirror neurons active during both observation and execution of actions, are thought to play a crucial role in imitation and other social-communicative skills that are often impaired in ASD (Oberman et al. 2008; Martineau et al. 2008; Hamilton 2013). Subjects with ASD show significant mu suppression to self-movements but they fail to react to the movements performed by others. These findings support the hypothesis of a dysfunctional MNS in individuals with ASD (Palau-Baduell et al. 2011; Southgate and Hamilton 2008; Leighton et al. 2008)

Oberman et al. (2008) reported that the MNS responded in individuals with autism, but only when observing a familiar face (Oberman et al. 2008). The same year, another author indicated that females showed stronger mu suppression than males, supporting the theory that autism represents an "extreme male brain" (Cheng et al. 2008).

A significant correlation between age and mu suppression in response only to the observation of actions (not during execution), for individuals with ASD and typical individuals was reported by Oberman et al. (2013). This result provides evidence against the argument that mirror neuron dysfunction improves with age in individuals with ASD, and suggests that a diagnosis-independent developmental change may be at the root of the correlation of age and mu suppression (Oberman et al. 2013).

To examine the integrity of the mirror system in autism Hamilton et al. (2013) systematically reviewed 25 suitable papers using neuroscience methods. They showed that the only well localized measure of mirror system function was functional magnetic resonance imaging (fMRI; Hamilton 2013).

The study of Ruysschaert et al. (2014) challenged the "broken mirror" hypothesis of ASD, suggesting that impaired neural mirroring is not a distinctive feature of ASD (Ruysschaert et al. 2014). Mu suppression was investigated in children with ASD during the observation of goal-directed actions and nongoal-directed mimicked hand movements, as well as during action execution. No significant correlations between mu suppression, quality of imitation, age, and social communication questionnaire scores were found by these authors. Overall, there is little evidence for a global dysfunction of the mirror system in autism.

Other available theories of the pathophysiology of ASD have focused on abnormal temporal coordination of neural activity in cortical circuits as a core impairment of the disorder. Synchronous neural oscillatory activity in the gamma range (30–80 Hz) has been shown to be abnormal in individuals with ASD and their first-degree relatives in response to simple auditory and language stimuli (McFadden et al. 2012; Milne et al. 2009). Magnetoencephalography (MEG) data also provides evidence that gamma-band activity may be involved in abnormal brain functioning in ASD. Limin Sun's findings highlighted the contribution of impaired gamma-band activity toward complex visual processing, suggesting atypical modulation of high-frequency power in fronto-posterior networks (Sun et al. 2012).

Lastly, an emerging focus of research on autism targets the identification of early-developing ASD endophenotypes using infant siblings of the affected children. Gabard-Durnam et al. (2013) demonstrated that low- and high-risk infants show different patterns of alpha asymmetry at 6 months of age and opposite growth trajectories in asymmetry over the following 12 months. These results support the candidacy of alpha asymmetry as an early neural ASD endophenotype (Gabard-Durnam et al. 2013; Carson et al. 2014).

4.2.1 Quantitative EEG

Analysis of resting state brain activity, using electrophysiological measures like complexity as well as functional connectivity, is of growing interest in the study of ASD, in this respect quantitative electroencephalography (QEEG) may help in detecting the regions of altered brain function and connectivity abnormalities (Billeci et al. 2013).

Children with ASD exhibit regionally specific elevations in delta, theta, alpha, and high-frequency (20–120 Hz) power, supporting an imbalance of neural excitation/inhibition as a neurobiological feature of this disorder. Moreover, increased temporal and parietal alpha power has been associated with greater symptom severity (Cornew et al. 2012).

Resting-state EEG studies of ASD suggest a U-shaped profile of electrophysiological power alterations, with excessive power in low-frequency and high-frequency bands, abnormal functional connectivity, and enhanced power in the left hemisphere of the brain. These studies have documented differences associated with ASD, particularly in frontal areas functionally linked to cognitive functions which are disrupted in individuals with ASD (Billeci et al. 2013; Wang et al. 2013; Shimizu et al. 1982).

Neuroimaging technologies and EEG studies have shown that autism is largely a disorder of neuronal connectivity (Coben et al. 2014; Palau-Baduell et al. 2012). Functional findings revealed that patients with ASD had deficit in long-distance connections (under-connectivity), with a most prominent deficit in fronto-posterior connections, as well as an excess of local connections (over-connectivity) and of long-distance under-connectivity, with some nonuniformities, along with disruptions were also described, more severe in later-developing cortical regions (Wass 2011; Palau-Baduell et al. 2012). Furthermore, lower synchronization in non-rapid eye movement (NREM) sleep stages, and low coherence for most frequency bands confirm the validity of the underconnectivity model in autism (Coben et al. 2008; Kulisek et al. 2008)

Recently, Barttfeld P. showed a decay in functional connectivity mainly within the delta and theta bands (the lower part of the EEG spectrum) associated with an increasing number of autistic traits. According to this author, EEG functional connectivity at low frequencies and its associated network properties may be related to some autistic features in the general population (Barttfeld et al. 2013). Resting state MEG has also revealed band-specific group differences in connectivity measure that agreed with other functional studies in fMRI and EEG (Ghanbari et al. 2013).

Cantor et al. (1986), were the first to examine the utility of pairwise coherence measures for representing connectivity impairments in autism (Cantor et al. 1986). Using phase coherence in multiple frequency EEG bands as a measure of functional connectivity, Pineda et al. (2012) have shown evidence for both global hypoconnectivity and local hyperconnectivity in individuals with ASD (Pineda et al. 2012). Carson et al. (2014) also found decreased interhemispheric connectivity in frontal and temporal-parietal regions in these children compared to controls using EEG coherence (Carson et al. 2014).

A large case control study conducted by Duffy et al. (2012) found a stable pattern of EEG spectral coherence distinguishing children with autism from neurotypical controls. They proposed that the predominantly reduced short-distance coherences may indicate poor local network function, whereas, the increased long-distance coherences might represent compensatory processes or reduced neural pruning (Duffy and Als 2012). It has been suggested that preferential attention to detail (perceptual domain) is connected with both, lower levels of alpha activity and reduced coherence in posterior regions in the ASD group (Mathewson et al. 2012).

EEG coherence has also been evaluated during the intermittent photic stimulation at fixed frequencies of 3–24 Hz in 14 boys with autism, aged 6–14 years, with relatively intact verbal and intellectual functions and without differences in the spontaneous EEG. However, the number of interhemispheric coherent connections pertaining to the 20 highest connections of each individual was significantly lower in autistic patients than in the controls at all the EEG beta frequencies corresponding to those of stimulation (Lazarev et al. 2013).

In this respect, there are a number of extant questions. First, whether aberrant connectivity observed in ASD should be seen as part of their primary pathogenesis, or whether it disrupted connectivity emerging over time. Second, how the patterns of disrupted connectivity found in ASD might relate to those found in a range of other disorders.

4.2.2 EEG Biofeedback as Treatment in ASD

Neurofeedback (NF) training is an intervention based on operant conditioning that results in self-regulation of brain electrical oscillations (Kubik 2010). NF exploits the brain's plasticity to normalize aberrant connectivity patterns apparent in the autistic brain, by grounding this training in known anatomical (e.g., mirror neuron system) and functional markers (e.g., mu rhythms) of autism. So, NF training holds promise to support current treatments for this complex disorder. Pineda et al. proposed a hypothesis which states that neurofeedback-induced alpha mu (8–12 Hz) rhythm suppression or desynchronization, a marker of cortical activation, should induce neuroplastic changes and lead to the normalization in relevant mirroring networks that have been associated with higher-order social cognition(Pineda et al. 2012).

Holtmann et al. (2011) reviewed available studies on the effectiveness of NF as a method of treatment for ASD core symptoms. The existing evidence at that time did not support the use of this tool in the treatment of ASD (Holtmann et al. 2011). Studies with outcomes in favor of NF might show an improvement in comorbid attention-deficit-hyperactivity disorder symptoms, rather than a true improvement in core ASD symptoms (Coben and Myers 2010; Holtmann et al. 2011; Kubik 2010).

Recently, a randomized pretest–posttest control group design showed that EEG biofeedback seems to be an applicable tool to regulate EEG activity, with specific effects on cognitive flexibility, but it did not result in significant reductions in symptoms of ASD (Kouijzer et al. 2013).

Despite conflicting results, literature analysis suggests that QEEG features are sensitive to modification in neuronal regulation dysfunction which characterizes autistic brain. The use of advanced techniques for the increase of the specificity and of spatial localization could allow finding distinctive patterns of QEEG abnormalities in ASD subjects, paving the way for the development of tailored intervention strategies.

4.3 Relationship Between EEG Epileptiform Abnormalities, Epilepsy, and Regression in ASD

4.3.1 EEG Epileptiform Abnormalities

Confounding interpretations of findings related to the co-occurrence of autism and epilepsy is that the standards used to determine epileptiform abnormalities (EA) in children with ASD have varied among studies (Tuchman et al. 2010). Both nonspecific changes, such as slowing or asymmetry, and epileptiform discharges, consisting of spikes or sharp wave discharges, sharp slow waves, generalized spike-wave, and generalized polyspikes are seen in this disorder (Spence and Schneider 2009). On the other hand, the term subclinical or nonconvulsive seizure is used to refer to electrographic patterns, without clinically recognizable cognitive, behavioral, or motor functions or any apparent impairment of consciousness (Chez et al. 2006).

Many authors have reported a prevalence of epileptiform abnormalities in 20–70 % of the children with autism and of epilepsy in 10–40 % of them (Rutter 1970; Tuchman 1997; Tuchman 2004; Rossi et al. 1995; Hashimoto et al. 2001; Hrdlicka et al. 2004; Hughes and Melyn 2005; Kawasaki et al. 1997; Ballaban-Gil and Tuchman 2000; Lewine et al. 1999). Moreover, high rates of interictal epileptiform EEG abnormalities in children with ASD have also been observed in patients with or without a history of seizures (Palau-Baduell et al. 2013; Chez et al. 2006; Kim et al.

2006; Parmeggiani et al. 2002; Amiet et al. 2008; Bolton et al. 2011; Francis et al. 2013). EA rates as high as 60 % have been seen even in the absence of epilepsy, so their presence should not be considered evidence of epilepsy. This raises questions about whether these discharges could be considered a biomarker of cortical dysfunction in this population, and whether these discharges have a causal association with any of the autism phenotypes. Some investigators propose that these abnormalities may play a causal role in the autism phenotype (Munoz-Yunta et al. 2003; Bolton et al. 2011; Dawson et al. 2005; Francis et al. 2013).

Kagan-Kushnir et al. (2005) summarized 13 EEG studies from 1966 to 2003. None of the studies received a "good" quality rating, three were rated "fair," and the other ten were rated "poor." In examining the prevalence of epileptiform abnormalities in all patients, irrespective of clinical seizure history, wide variants in prevalence rates were obtained, ranging from 10.3 to 72.4 %. When only the "fair" studies were considered, the rates ranged from 38.3 to 60.8 %. Out of 13, 8 studies examined the prevalence rate of epileptiform abnormalities in patients with no clinical history of seizures and it was found to range from 6.1 to 31 % (Kagan-Kushnir et al. 2005).

In the general population the occurrence of EA observed ranges from 2 to 8.7 %, and decreases during puberty (Danielsson et al. 2005; Hara 2007; Trevathan 2004), while EA occurrence in ASD ranges from 6.7 to 83.0 %, tends to disappear during puberty (Parmeggiani et al. 2007), and is present predominantly in female subjects (Chez et al. 2006). On the other hand, the prevalence of epilepsy declines with age in individuals with static processes like cerebral palsy and intellectual disability, nevertheless increases of autism and epilepsy raise the possibility of an ongoing degenerative process in these individuals, like the occurrences in Down syndrome (Silver and Rapin 2012; Tuchman et al. 2010). Epilepsy generally persists into adult life, however a remission of only 16 % has been reported in adults with autism (Levisohn 2007). The heterogeneous data reported in the literature are probably due to (a) different samples with different ages and clinical features and (b) the methodological variability in collecting and interpreting EEGs.

It has also been demonstrated that the occurrence of EA is higher when using 24- to 48-h EEG recordings than in routine studies; in fact, this may be present only during sleep (Parmeggiani et al. 2010; Hrdlicka et al. 2004). According to one report EEG sleep study, could be best achieved in patients with autism by using dexmedetomidine, (Ray and Tobias 2008).

MEG and sleep EEG are more sensitive for correctly detecting epileptiform EEG abnormalities in autism than wake EEGs (Lewine et al. 1999; Hrdlicka 2008). Mulligan et al. (2014) completed a retrospective chart review of 101 patients with ASD who had overnight EEGs they suggested that increasing severity of autistic symptoms may be associated with a higher likelihood of epileptiform abnormalities (Mulligan and Trauner 2014). MEG epileptiform activity is frequently documented in children with early-onset ASD and subclinical epileptiform activity is present especially in the perisylvian regions for many patients with this pathology (Munoz-Yunta et al. 2008). The focus of spike discharges has been reported to be mainly in the centro-temporal or temporal (Olsson et al. 1988; Tuchman and Rapin 1997;

Parmeggiani et al. 2010), frontal (Kawasaki et al. 1997; Hashimoto et al. 2001), or rarely in the occipital regions (Nass et al. 1998). Kanemura et al. (2013) observed that the presence of frontal paroxysms was significantly associated with the later development of epilepsy compared with centrotemporal paroxysmus (Kanemura et al. 2013).

In a recent study, our team in the International Center for Neurological Restoration in Havana Cuba evaluated 70 ASD patients between the age of 2 and 14 years (5.8 ± 2.86) with male predominance (63.3 %), 55 % of the patients were classified as primary ASD. Sleep EEGs, wake EEGs, or both were performed in this study. Sleep EEGs recording was performed in 41.6 % of the patients. The overall rate of epileptiform EEG abnormalities in the whole sample was 71.6 %. These EEG findings were not found to be associated with clinical severity of autism. No significant relationship between EA occurrence, and ASD subtypes was observed (Fisher exact test $p = 0.46$). Epilepsy was diagnosed in 11, 6 %, of the study participants. Temporal lobe localization of the EA was associated with epilepsy diagnosis, (unpublished data).

In general, the significance of EA is controversial. Moreover, a clinical diagnosis of ASD cannot be confirmed by pathognomic EEG or PSG features but these data may help to determine the role of abnormal electrical activity in the development of ASD.

4.3.2 EEG Epileptiform Abnormalities and Epilepsy

Epilepsy is the most common neurological comorbidity in autism. Approximately one third of the children with autism develop epilepsy (Giovanardi et al. 2000; Olsson et al. 1988), with prevalence estimates between 8 and 42 % (Volkmar and Nelson 1990; Rossi et al. 1995; Spence and Schneider 2009; Tuchman and Rapin 2002; Hughes and Melyn 2005; Kagan-Kushnir et al. 2005; Danielsson et al. 2005; Francis et al. 2013).

The onset of epilepsy in autism has two peaks: one before 5 years of age and the other after 10–12 years of age, with most cases presenting after 10 years of age (Volkmar et al. 2005). In these cases all types of epilepsy have been observed (Volkmar et al. 2005; Gillberg 1991).

The high frequency of autism in some of the early-onset developmental encephalopathic epilepsies, and the high prevalence of interictal EEG discharges in children with autism is frequently cited as evidence of the relationship between autism and epilepsy (Berg and Plioplys 2012).

Some studies have ascertained that mental retardation, cerebral lesions, and rare diseases associated with ASD increase the risk of epilepsy (Spence and Schneider 2009; Hughes and Melyn 2005). The prevalence of epilepsy reported in autism has varied across studies depending on the age distribution of the sample, the degree of mental retardation, and the type of language disorder (Volkmar et al. 2005; Kagan-Kushnir et al. 2005). However, in "idiopathic" ASD seizure the occurrence remains higher than in the general population, suggesting that autism itself is associated with an enhanced risk of epilepsy (Spence and Schneider 2009). A paper published by Ekinci et al. (2010) examined the possible associations of epilepsy/interictal epileptiform abnormalities with asthma/allergy, hyperactivity, and familial factors in ASD. They observed that epilepsy was associated with a family history of epilepsy and psychiatric problems in the mother during pregnancy (Ekinci et al. 2010).

It has been reported that some ASD patients with late-onset epilepsy showed severe EEG abnormalities, which included continuous spike-waves during slow-wave sleep (CSWS), generally demonstrate an improvement in EEG and clinical symptoms in the long-term follow up (Lee et al. 2011).

The complexity of the relationship of ASD, epilepsy, and epileptiform EEG activity is highlighted in tuberous sclerosis complex (TSC). This is a key clinical model for at least four reasons. First, 1–5% of the children with autism have TSC (de et al. 2005). Second, a more careful analysis reveals that TSC is present in 8–14% of those with the autism–epilepsy phenotype (Smalley 1998).Third, autism or ASD have been reported in up to 50% of the individuals with TSC (de et al. 2005). Fourth, approximately 60% of the children with TSC have epilepsy and 50% have infantile spasms (Guerrini and Aicardi 2003). All of these reasons make TSC a unique clinical model to study the complex interplay between genetics, seizures onset, and location of epileptiform activity to the development of ASD.

The consensus emerging from studies on ASD and epilepsy is that the same brain pathology accounts for the majority of children with co-occurring ASD and epilepsy or with an epileptiform EEG. This brain pathology may represent a set of uniform underlying genetics as well. The current understanding of the association between epilepsy and ASD is still limited, but from a clinical point of view, this association should not be overlooked. Controversy also seems to exist regarding the rate of seizures and EEG abnormalities.

4.3.3 EEG Epileptiform Abnormalities and Autistic Regression

Another interesting point emerging from studying the association of epilepsy, EA, and autism is the fact of autistic regression (usually occurring between 18 and 24 months of age) in which the developmental trajectory of approximately 30% of the children with ASD is characterized by a regression of verbal and nonverbal communication skills. This phenomenon is currently well accepted, but poorly understood at a biological level (Werner et al. 2005; Luyster et al. 2005; Lord et al. 2004; Goldberg et al. 2003; Meilleur and Fombonne 2009; Baird et al. 2008).

Many studies have addressed the question of the influence of epilepsy and/or EA on autistic regression (Tuchman 1997; Kobayashi and Murata 1998; Hrdlicka et al. 2004; Baird et al. 2006; Deonna and Roulet 2006). However, the relationship between them continues to be an area of active research interest and controversy with studies showing mixed results. Some studies during the past decade have found no differences in history of autistic regression in ASD children with epileptiform

EEGs and epilepsy versus ASD children with a normal EEG and no epilepsy (Hara 2007; Canitano et al. 2005; Hrdlicka 2008; Parmeggiani et al. 2010). Contrary to results showing no relationship of regression to epilepsy in autism there are other case reports linking epilepsy or an epileptiform EEG to autistic regression (Deonna and Roulet 2006; Chilosi et al. 2013; Hrdlicka et al. 2004; Giannotti et al. 2008). This latter study also found that children with autistic regression had more disrupted sleep than those with ASD without regression (Giannotti et al. 2008).

For many years, a number of studies have pointed out the role of EEG abnormalities in language and cognitive processes suggesting that interictal epileptiform activity could impair brain function (Patry and Naquet 1971; Deonna 1996; Rapin 1996; Aarts et al. 1984; Binnie et al. 1987; Kasteleijn-Nolst Trenite et al. 1987) while, uncontrolled and few randomized controlled trials of antiepileptic treatment of interictal epileptiform discharges have suggested that the suppression of discharges is associated with significant improvement in psychosocial function (Garcia-Penas 2011).

Furthermore, despite the high prevalence of interictal EEG epileptiform activity in children with ASD and the overlap of Landau–Kleffner syndrome (LKS) and ASD, little evidence to date has shown that spikes contribute to the pathogenesis or to the worsening of language, social, or behavioral dysfunction in children with ASD (Tharp 2004). However, there are clear differences between LKS and autistic regression (Szatmari et al. 2008).

An attempt to clarify the relationship between regressions or any type of problematic clinical findings and severity of EEG abnormalities in ASD patients with late-onset epilepsy was made by Lee et al. (2011). They found that severe EEG abnormalities tended to be related to the neuropsychological function. Nevertheless they were not able to conclude that the treatment of EEG abnormality was related to the improved neuropsychological symptoms (Lee et al. 2011).

In summary, although there is no clear understand on the biology of regression in autism, we have much information of their phenomenology. The decision to treat EA remains controversial, and the treatment of seizures should be pursued utilizing drugs generally indicated in international epilepsy guidelines. It is quite evident that the results of different studies are argumentative and the relationship of epilepsy and EEG abnormalities to developmental regression in autism remains unknown and should be clarified and investigated.

4.4 Sleep Disorders and Polysomnography in ASD

Just as with epileptiform discharges and epilepsy, the prevalence of sleep problems in children with ASD is very high, ranging between 40 and 80 % (Thirumalai et al. 2002; Silver and Rapin 2012; Richdale 1999; Schreck and Mulick 2000; Wiggs and Stores 2004; Polimeni et al. 2005; Liu et al. 2006; Malow 2004; Malow et al. 2009; Bruni et al. 2007; Goldman et al. 2009). A growing body of literature contains reports of sleep disorders as a third indicator of abnormal neural functioning in autism and, therefore, a characteristic of the ASD phenotype (Anders et al. 2011; Wiggs and Stores 2004; Richdale 1999; Cortesi et al. 2010; Baker et al. 2013). Parental surveys indicate a 50–80 % prevalence of sleep problems in children with ASD, compared with a 9–50 % prevalence rate in age-matched, typically developing subjects (Polimeni et al. 2005; Kotagal and Broomall 2012; Allik et al. 2006; Richdale and Schreck 2009).

The degree of cognitive impairment likely does not influence the prevalence of sleep problems in ASD because they are observed in those who are severely mentally handicapped as well as in those who are high functioning, with intelligence quotients greater than 70 (Richdale and Schreck 2009; Krakowiak et al. 2008). Sleep problems also do not seem to be influenced by ASD subtype (Polimeni et al. 2005). A positive correlation has also been observed between the number of seizures and total scores on the Child Sleep Habits Questionnaire, where higher scores denote more sleep problems (Giannotti et al. 2008). Parents of children with ASD and epilepsy are more likely to report difficulties with sleep than parents of children without epilepsy (Cotton and Richdale 2006; Goldman et al. 2012; Wiggs and Stores 2004).

The main problems described in ASD include insomnia, difficulty with sleep initiation and sleep maintenance, parasomnias such as rapid eye movement (REM) and non-REM arousal disorders, rhythmical movement disorders, and periodic limb movements during sleep (Malow et al. 2006; Liu et al. 2006; Krakowiak et al. 2008; Hoshino et al. 1984). The multiplicity of sleep problems in children with ASDs was confirmed by Wiggs and Stores using sleep diaries from 69 parents, together with actigraphy (a device to detect and record muscle activity), to monitor sleep pattern (Wiggs and Stores 2004). Although some sleep disorders might be of behavioral origin (as in the typically developing children with poor sleep hygiene), circadian rhythm disturbances have also been reported in children with ASD (Patzold et al. 1998; Wiggs and Stores 2004).

The consequences of the difficulties with sleep initiation and maintenance in children with ASD may include alterations in daytime behavior, memory, learning, and also significant stress in caretakers (Kotagal and Broomall 2012). In 2004 Schreck et al. demonstrated that in children with autism, fewer hours of sleep per night were highly correlated with a more severe behavioral phenotype (Schreck et al. 2004). Previous studies had also suggested that some symptoms specific to autism may be directly associated with disturbed sleep, and that the improved sleep in children with autism is associated with enhancement in daytime behavior (Johnson et al. 1996; Patzold et al. 1998; Richdale 1999; Segawa and Nomura 1992).

In an interesting study Giannotti et al. (2006) observed persistent sleep problems in children with autism, in over 50 % of the sample (Giannotti et al. 2006), with a peak age of onset during the second year of life, similar to what is known for regression. Later, this author also reported significantly more severe and persistent abnormalities of sleep wake patterns in regressed children (Giannotti et al. 2008; Giannotti et al. 2011).

Polysomnography (PSG) consists of monitoring multiple simultaneous, physiologic parameters during sleep Fig. 4.1 and it is indicated when sleep-disordered

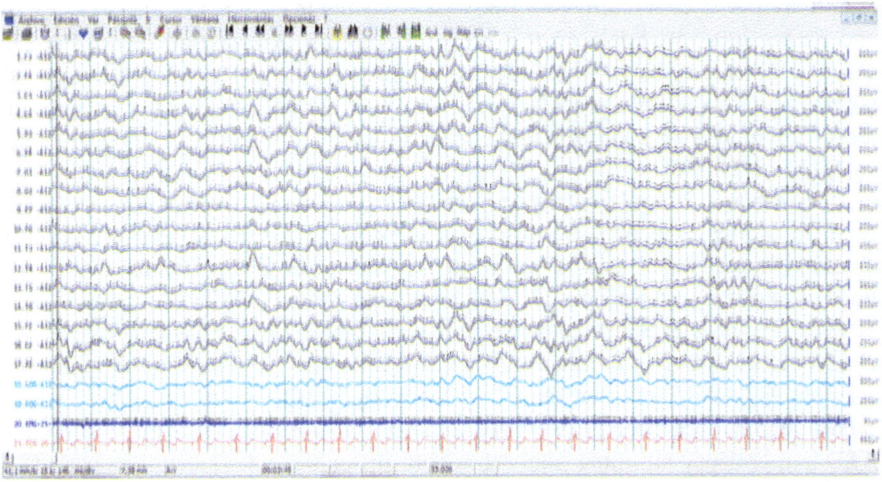

Fig. 4.1 Polysomnography study in a 4-year-old patient with ASD without epileptic seizures. Recordings involve complete electroencephalogram (EEG), chin electromyography, eye movements and electrocardiogram. Note EEG epileptiform abnormality in channels containing the right centroparietotemporal leads, C4, P4, T4, and T6

breathing, restless legs syndrome, parasomnias, or nocturnal seizures are suspected. PSG studies have confirmed the presence of disrupted sleep architecture in children with ASD. PSG abnormalities include reduction of REM sleep; longer sleep latency, increased arousals, lower sleep efficiency, increased stage 1 sleep, and decreased slow wave sleep as well as decreased density of spindle activity (Palau-Baduell et al. 2013).

Relatively few reports of PSG-based sleep studies in children with autism have been published. Overall, these have focused on abnormalities in REM sleep, including immaturity in the organization of eye movements into discrete bursts (Tanguay et al. 1976), increased muscle twitches during REM sleep (Elia et al. 2000), and undifferentiated sleep in which features of non-REM and REM sleep are intermixed (Diomedi et al. 1999; Limoges et al. 2005). The prolonged sleep times, early wake times, and frequent interruptions in sleep noted in the survey literature have not been commented on in these PSG studies. Although subjective sleep parameters appear to be roughly similar in adults and children with ASD, there are inconsistencies in the objective sleep profiles obtained from actual sleep recordings (Elia et al. 2000).

The characterization of sleep manifestations by specific ASD subtype (preferably by using the International Classification of Sleep Disorders) will help gather better longitudinal data about diagnosis and responses to specific treatments.

4.5 Conclusions

EEG and MEG provide evidence of disrupted brain connectivity in ASD and reveal that gamma-band activity may be crucially involved in aberrant brain functioning in ASD. However, researchers have only recently begun to link patterns of brain activity and connectivity to behavior.

Although there have been relatively few reports of PSG-based sleep studies in ASD, both non-REM and REM sleep abnormalities have been observed, which support the multiplicity of sleep problems in this pathology.

One of the best-known associations with central nervous system dysfunction in ASD is the high risk of epilepsy; though the relationship among epileptiform abnormalities, epilepsy, and regression is not yet well understood. Thus, a greater number of controlled studies are required in order to confirm this relationship. That is why, identification of the early processes that leads to ASD, and to epileptogenicity, or both is a challenge that is worth pursuing.

Acknowledgments We thank Odalys Morales Chacón for her English assistance. We would also like to thank Abel Sanchez, Daymet Grass, Maydelin Alfonso, and Maria Luisa Rodriguez for their useful cooperation.

References

Aarts JH, Binnie CD, Smit AM, Wilkins AJ (1984) Selective cognitive impairment during focal and generalized epileptiform EEG activity. Brain 107:293–308

Allik H, Larsson JO, Smedje H (2006) Sleep patterns of school-age children with Asperger syndrome or high-functioning autism. J Autism Dev Disord 36:585–595

Amiet C, Gourfinkel-An I, Bouzamondo A, Tordjman S, Baulac M, Lechat P, Mottron L, Cohen D (2008) Epilepsy in autism is associated with intellectual disability and gender: evidence from a meta-analysis. Biol Psychiatry 64:577–582

Anders TF, Iosif AM, Schwichtenberg AJ, Tang K, Goodlin-Jones BL (2011) Six-month sleep-wake organization and stability in preschool-age children with autism, developmental delay, and typical development. Behav Sleep Med 9:92–106

Baird G, Robinson RO, Boyd S, Charman T (2006) Sleep electroencephalograms in young children with autism with and without regression. Dev Med Child Neurol 48:604–608

Baird G, Charman T, Pickles A, Chandler S, Loucas T, Meldrum D, Carcani-Rathwell I, Serkana D, Simonoff E (2008) Regression, developmental trajectory and associated problems in disorders in the autism spectrum: the SNAP study. J Autism Dev Disord 38:1827–1836

Baker E, Richdale A, Short M, Gradisar M (2013) An investigation of sleep patterns in adolescents with high-functioning autism spectrum disorder compared with typically developing adolescents. Dev Neurorehabil 16:155–165

Ballaban-Gil K, Tuchman R (2000) Epilepsy and epileptiform EEG: association with autism and language disorders. Ment Retard Dev Disabil Res Rev 6:300–308

Barttfeld P, Amoruso L, Ais J, Cukier S, Bavassi L, Tomio A, Manes F, Ibanez A, Sigman M (2013) Organization of brain networks governed by long-range connections index autistic traits in the general population. J Neurodev Disord 5:16–5

Berg AT, Plioplys S (2012) Epilepsy and autism: is there a special relationship? Epilepsy Behav 23:193–198

Billeci L, Sicca F, Maharatna K, Apicella F, Narzisi A, Campatelli G, Calderoni S, Pioggia G, Muratori F (2013) On the application of quantitative EEG for characterizing autistic brain: a systematic review. Front Hum Neurosci 7:442. doi:10.3389/fnhum.2013.00442. eCollection;%2013.:442

Binnie CD, Kasteleijn-Nolst Trenite DG, Smit AM, Wilkins AJ (1987) Interactions of epileptiform EEG discharges and cognition. Epilepsy Res 1:239–245

Bolton PF, Carcani-Rathwell I, Hutton J, Goode S, Howlin P, Rutter M (2011) Epilepsy in autism: features and correlates. Br J Psychiatry 198:289–294

Bruni O, Ferri R, Vittori E, Novelli L, Vignati M, Porfirio MC, Arico D, Bernabei P, Curatolo P (2007) Sleep architecture and NREM alterations in children and adolescents with Asperger syndrome. Sleep 30:1577–1585

Canitano R, Luchetti A, Zappella M (2005) Epilepsy, electroencephalographic abnormalities, and regression in children with autism. J Child Neurol 20:27–31

Cantor DS, Thatcher RW, Hrybyk M, Kaye H (1986) Computerized EEG analyses of autistic children. J Autism Dev Disord 16:169–187

Carson AM, Salowitz NM, Scheidt RA, Dolan BK, Van Hecke AV (2014) Electroencephalogram coherence in children with and without autism spectrum disorders: decreased interhemispheric connectivity in autism. Autism Res10

Cheng Y, Lee PL, Yang CY, Lin CP, Hung D, Decety J (2008) Gender differences in the mu rhythm of the human mirror-neuron system. PLoS ONE 3:e2113

Chez MG, Chang M, Krasne V, Coughlan C, Kominsky M, Schwartz A (2006) Frequency of epileptiform EEG abnormalities in a sequential screening of autistic patients with no known clinical epilepsy from 1996 to 2005. Epilepsy Behav 8:267–271

Chilosi AM, Brovedani P, Ferrari AR, Ziegler AL, Guerrini R, Deonna T (2013) Language regression associated with autistic regression and electroencephalographic (EEG) abnormalities: a prospective study. J Child Neurol

Coben R, Myers TE (2010) The relative efficacy of connectivity guided and symptom based EEG biofeedback for autistic disorders. Appl Psychophysiol Biofeedback 35:13–23

Coben R, Clarke AR, Hudspeth W, Barry RJ (2008) EEG power and coherence in autistic spectrum disorder. Clin Neurophysiol 119:1002–1009

Coben R, Mohammad-Rezazadeh I, Cannon RL (2014) Using quantitative and analytic EEG methods in the understanding of connectivity in autism spectrum disorders: a theory of mixed over- and under-connectivity. Front Hum Neurosci 8:45. doi:10.3389/fnhum.2014.00045. eCollection;%2014

Cornew L, Roberts TP, Blaskey L, Edgar JC (2012) Resting-state oscillatory activity in autism spectrum disorders. J Autism Dev Disord 42:1884–1894

Cortesi F, Giannotti F, Ivanenko A, Johnson K (2010) Sleep in children with autistic spectrum disorder. Sleep Med 11:659–664

Cotton S, Richdale A (2006) Brief report: parental descriptions of sleep problems in children with autism, Down syndrome, and Prader-Willi syndrome. Res Dev Disabil 27:151–161

Danielsson S, Gillberg IC, Billstedt E, Gillberg C, Olsson I (2005) Epilepsy in young adults with autism: a prospective population-based follow-up study of 120 individuals diagnosed in childhood. Epilepsia 46:918–923

Dawson G, Webb SJ, Wijsman E, Schellenberg G, Estes A, Munson J, Faja S (2005) Neurocognitive and electrophysiological evidence of altered face processing in parents of children with autism: implications for a model of abnormal development of social brain circuitry in autism. Dev Psychopathol 17:679–697

de VP, Humphrey A, McCartney D, Prather P, Bolton P, Hunt A (2005) Consensus clinical guidelines for the assessment of cognitive and behavioural problems in Tuberous Sclerosis. Eur Child Adolesc Psychiatry 14:183–190

Deonna T (1996) Autism and related conditions. J Am Acad Child Adolesc Psychiatry 35:403–405

Deonna T, Roulet E (2006) Autistic spectrum disorder: evaluating a possible contributing or causal role of epilepsy. Epilepsia 47(Suppl 2):79–82

Diomedi M, Curatolo P, Scalise A, Placidi F, Caretto F, Gigli GL (1999) Sleep abnormalities in mentally retarded autistic subjects: Down's syndrome with mental retardation and normal subjects. Brain Dev 21:548–553
Duffy FH, Als H (2012) A stable pattern of EEG spectral coherence distinguishes children with autism from neuro-typical controls—a large case control study. BMC Med 10:64. doi:10.1186/1741-7015-10-64.:64-10
Ekinci O, Arman AR, Isik U, Bez Y, Berkem M (2010) EEG abnormalities and epilepsy in autistic spectrum disorders: clinical and familial correlates. Epilepsy Behav 17:178–182
Elia M, Ferri R, Musumeci SA, Del GS, Bottitta M, Scuderi C, Miano G, Panerai S, Bertrand T, Grubar JC (2000) Sleep in subjects with autistic disorder: a neurophysiological and psychological study. Brain Dev 22:88–92
Francis A, Msall M, Obringer E, Kelley K (2013) Children with autism spectrum disorder and epilepsy. Pediatr Ann 42:255–260
Gabard-Durnam L, Tierney AL, Vogel-Farley V, Tager-Flusberg H, Nelson CA (2013) Alpha asymmetry in infants at risk for autism spectrum disorders. J Autism Dev Disord 45(2): 473–480. doi:10.1007/s10803-013-1926-4
Garcia-Penas JJ (2011) Interictal epileptiform discharges and cognitive impairment in children. Rev Neurol 52(Suppl 1):S43–S52
Ghanbari Y, Bloy L, Christopher EJ, Blaskey L, Verma R, Roberts TP (2013) Joint analysis of band-specific functional connectivity and signal complexity in autism. J Autism Dev Disord 45(2):444–460. doi:10.1007/s10803-013-1915-7
Giannotti F, Cortesi F, Cerquiglini A, Bernabei P (2006) An open-label study of controlled-release melatonin in treatment of sleep disorders in children with autism. J Autism Dev Disord 36:741–752
Giannotti F, Cortesi F, Cerquiglini A, Miraglia D, Vagnoni C, Sebastiani T, Bernabei P (2008) An investigation of sleep characteristics, EEG abnormalities and epilepsy in developmentally regressed and non-regressed children with autism. J Autism Dev Disord 38:1888–1897
Giannotti F, Cortesi F, Cerquiglini A, Vagnoni C, Valente D (2011) Sleep in children with autism with and without autistic regression. J Sleep Res 20:338–347
Gillberg C (1991) The treatment of epilepsy in autism. J Autism Dev Disord 21:61–77
Giovanardi RP, Posar A, Parmeggiani A (2000) Epilepsy in adolescents and young adults with autistic disorder. Brain Dev 22:102–106
Goldberg WA, Osann K, Filipek PA, Laulhere T, Jarvis K, Modahl C, Flodman P, Spence MA (2003) Language and other regression: assessment and timing. J Autism Dev Disord 33:607–616
Goldman SE, Surdyka K, Cuevas R, Adkins K, Wang L, Malow BA (2009) Defining the sleep phenotype in children with autism. Dev Neuropsychol 34:560–573
Goldman SE, Richdale AL, Clemons T, Malow BA (2012) Parental sleep concerns in autism spectrum disorders: variations from childhood to adolescence. J Autism Dev Disord 42:531–538
Guerrini R, Aicardi J (2003) Epileptic encephalopathies with myoclonic seizures in infants and children (severe myoclonic epilepsy and myoclonic-astatic epilepsy). J Clin Neurophysiol 20:449–461
Hamilton AF (2013) Reflecting on the mirror neuron system in autism: a systematic review of current theories. Dev Cogn Neurosci 3:91–105. doi:10.1016/j.dcn.2012.09.008. Epub 2012 Oct 13.
Hara H (2007) Autism and epilepsy: a retrospective follow-up study. Brain Dev 29:486–490
Hashimoto T, Sasaki M, Sugai K, Hanaoka S, Fukumizu M, Kato T (2001) Paroxysmal discharges on EEG in young autistic patients are frequent in frontal regions. J Med Invest 48:175–180
Holtmann M, Steiner S, Hohmann S, Poustka L, Banaschewski T, Bolte S (2011) Neurofeedback in autism spectrum disorders. Dev Med Child Neurol 53:986–993
Hoshino Y, Watanabe H, Yashima Y, Kaneko M, Kumashiro H (1984) An investigation on sleep disturbance of autistic children. Folia Psychiatr Neurol Jpn 38:45–51
Hrdlicka M (2008) EEG abnormalities, epilepsy and regression in autism: a review. Neuro Endocrinol Lett 29:405–409

Hrdlicka M, Komarek V, Propper L, Kulisek R, Zumrova A, Faladova L, Havlovicova M, Sedlacek Z, Blatny M, Urbanek T (2004) Not EEG abnormalities but epilepsy is associated with autistic regression and mental functioning in childhood autism. Eur Child Adolesc Psychiatry 13:209–213

Hughes JR (2010) A review of Savant syndrome and its possible relationship to epilepsy. Epilepsy Behav 17:147–152

Hughes JR, Melyn M (2005) EEG and seizures in autistic children and adolescents: further findings with therapeutic implications. Clin EEG Neurosci 36:15–20

Jeste SS (2011) The neurology of autism spectrum disorders. Curr Opin Neurol 24:132–139

Johnson CD, Matt MK, Dennison D, Brown RS, Koh S (1996) Preventing factitious gingival injury in an autistic patient. J Am Dent Assoc 127:244–247

Kagan-Kushnir T, Roberts SW, Snead OC III (2005) Screening electroencephalograms in autism spectrum disorders: evidence-based guideline. J Child Neurol 20:197–206

Kanemura H, Sano F, Tando T, Sugita K, Aihara M (2013) Can EEG characteristics predict development of epilepsy in autistic children? Eur J Paediatr Neurol 17:232–237

Kasteleijn-Nolst Trenite DG, Riemersma JB, Binnie CD, Smit AM, Meinardi H (1987) The influence of subclinical epileptiform EEG discharges on driving behaviour. Electroencephalogr Clin Neurophysiol 67:167–170

Kawasaki Y, Yokota K, Shinomiya M, Shimizu Y, Niwa S (1997) Brief report: electroencephalographic paroxysmal activities in the frontal area emerged in middle childhood and during adolescence in a follow-up study of autism. J Autism Dev Disord 27:605–620

Kelly PA (2014) Textual Standardization and the DSM-5 "Common Language". J Med Humanit 35(2):171–189. doi:10.1007/s10912-014-9281-9

Kim HL, Donnelly JH, Tournay AE, Book TM, Filipek P (2006) Absence of seizures despite high prevalence of epileptiform EEG abnormalities in children with autism monitored in a tertiary care center. Epilepsia 47:394–398

Kobayashi R, Murata T (1998) Setback phenomenon in autism and long-term prognosis. Acta Psychiatr Scand 98:296–303

Kotagal S, Broomall E (2012) Sleep in children with autism spectrum disorder. Pediatr Neurol 47:242–251

Kouijzer ME, van Schie HT, Gerrits BJ, Buitelaar JK, de Moor JM (2013) Is EEG-biofeedback an effective treatment in autism spectrum disorders? A randomized controlled trial. Appl Psychophysiol Biofeedback 38:17–28

Krakowiak P, Goodlin-Jones B, Hertz-Picciotto I, Croen LA, Hansen RL (2008) Sleep problems in children with autism spectrum disorders, developmental delays, and typical development: a population-based study. J Sleep Res 17:197–206

Kubik A (2010) Neurofeedback and personal development. Przegl Lek 67:716–720

Kulisek R, Hrncir Z, Hrdlicka M, Faladova L, Sterbova K, Krsek P, Vymlatilova E, Palus M, Zumrova A, Komarek V (2008) Nonlinear analysis of the sleep EEG in children with pervasive developmental disorder. Neuro Endocrinol Lett 29:512–517

Lazarev VV, Pontes A, Mitrofanov AA, Deazevedo LC (2013) Reduced interhemispheric connectivity in childhood autism detected by electroencephalographic photic driving coherence. J Autism Dev Disord 45(2):537–547. doi:10.1007/s10803-013-1959-8

Lee H, Kang HC, Kim SW, Kim YK, Chung HJ (2011) Characteristics of late-onset epilepsy and EEG findings in children with autism spectrum disorders. Korean J Pediatr 54:22–28

Leighton J, Bird G, Charman T, Heyes C (2008) Weak imitative performance is not due to a functional 'mirroring' deficit in adults with autism spectrum disorders. Neuropsychologia 46:1041–1049

Levisohn PM (2007) The autism-epilepsy connection. Epilepsia 48(Suppl 9):33–35

Lewine JD, Andrews R, Chez M, Patil AA, Devinsky O, Smith M, Kanner A, Davis JT, Funke M, Jones G, Chong B, Provencal S, Weisend M, Lee RR, Orrison WW Jr (1999) Magnetoencephalographic patterns of epileptiform activity in children with regressive autism spectrum disorders. Pediatrics 104:405–418

Limoges E, Mottron L, Bolduc C, Berthiaume C, Godbout R (2005) Atypical sleep architecture and the autism phenotype. Brain 128:1049–1061

Liu X, Hubbard JA, Fabes RA, Adam JB (2006) Sleep disturbances and correlates of children with autism spectrum disorders. Child Psychiatry Hum Dev 37:179–191

Lord C, Shulman C, DiLavore P (2004) Regression and word loss in autistic spectrum disorders. J Child Psychol Psychiatry 45:936–955

Luyster R, Richler J, Risi S, Hsu WL, Dawson G, Bernier R, Dunn M, Hepburn S, Hyman SL, McMahon WM, Goudie-Nice J, Minshew N, Rogers S, Sigman M, Spence MA, Goldberg WA, Tager-Flusberg H, Volkmar FR, Lord C (2005) Early regression in social communication in autism spectrum disorders: a CPEA Study. Dev Neuropsychol 27:311–336

Malow BA (2004) Sleep disorders, epilepsy, and autism. Ment Retard Dev Disabil Res Rev 10:122–125

Malow BA, Marzec ML, McGrew SG, Wang L, Henderson LM, Stone WL (2006) Characterizing sleep in children with autism spectrum disorders: a multidimensional approach. Sleep 29:1563–1571

Malow BA, Crowe C, Henderson L, McGrew SG, Wang L, Song Y, Stone WL (2009) A sleep habits questionnaire for children with autism spectrum disorders. J Child Neurol 24:19–24

Martineau J, Cochin S, Magne R, Barthelemy C (2008) Impaired cortical activation in autistic children: is the mirror neuron system involved? Int J Psychophysiol 68:35–40

Mathewson KJ, Jetha MK, Drmic IE, Bryson SE, Goldberg JO, Schmidt LA (2012) Regional EEG alpha power, coherence, and behavioral symptomatology in autism spectrum disorder. Clin Neurophysiol 123:1798–1809

McFadden KL, Hepburn S, Winterrowd E, Schmidt GL, Rojas DC (2012) Abnormalities in gamma-band responses to language stimuli in first-degree relatives of children with autism spectrum disorder: an MEG study. BMC Psychiatry 12:213. doi:10.1186/1471-244X-12-213

Meilleur AA, Fombonne E (2009) Regression of language and non-language skills in pervasive developmental disorders. J Intellect Disabil Res 53:115–124

Milne E, Scope A, Pascalis O, Buckley D, Makeig S (2009) Independent component analysis reveals atypical electroencephalographic activity during visual perception in individuals with autism. Biol Psychiatry 65:22–30

Mulligan CK, Trauner DA (2014) Incidence and behavioral correlates of epileptiform abnormalities in autism spectrum disorders. J Autism Dev Disord 44:452–458

Munoz-Yunta JA, Salvado B, Ortiz-Alonso T, Amo C, Fernandez-Lucas A, Maestu F, Palau-Baduell M (2003) Clinical features of epilepsy in autism spectrum disorders. Rev Neurol 36(Suppl 1):S61–S67

Munoz-Yunta JA, Ortiz T, Palau-Baduell M, Martin-Munoz L, Salvado-Salvado B, Valls-Santasusana A, Perich-Alsina J, Cristobal I, Fernandez A, Maestu F, Dursteler C (2008) Magnetoencephalographic pattern of epileptiform activity in children with early-onset autism spectrum disorders. Clin Neurophysiol 119:626–634

Nass R, Gross A, Devinsky O (1998) Autism and autistic epileptiform regression with occipital spikes. Dev Med Child Neurol 40:453–458

Oberman LM, Ramachandran VS, Pineda JA (2008) Modulation of mu suppression in children with autism spectrum disorders in response to familiar or unfamiliar stimuli: the mirror neuron hypothesis. Neuropsychologia 46:1558–1565

Oberman LM, McCleery JP, Hubbard EM, Bernier R, Wiersema JR, Raymaekers R, Pineda JA (2013) Developmental changes in mu suppression to observed and executed actions in autism spectrum disorders. Soc Cogn Affect Neurosci 8:300–304

Olsson I, Steffenburg S, Gillberg C (1988) Epilepsy in autism and autisticlike conditions. A population-based study. Arch Neurol 45:666–668

Palau-Baduell M, Valls-Santasusana A, Salvado-Salvado B (2011) Autism spectrum disorders and mu rhythm. A new neurophysiological view. Rev Neurol 52(Suppl 1):S141–S146

Palau-Baduell M, Salvado-Salvado B, Clofent-Torrento M, Valls-Santasusana A (2012) Autism and neural connectivity. Rev Neurol 54(Suppl 1):S31–S39

Palau-Baduell M, Valls-Santasusana A, Salvado-Salvado B, Clofent-Torrento M (2013) Interest of electroencephalogram in autism. Rev Neurol 56(Suppl 1):S35–S43
Parmeggiani A, Posar A, Giovanardi-Rossi P, Andermann F, Zifkin B (2002) Autism, macrocrania and epilepsy: how are they linked? Brain Dev 24:296–299
Parmeggiani A, Posar A, Antolini C, Scaduto MC, Santucci M, Giovanardi-Rossi P (2007) Epilepsy in patients with pervasive developmental disorder not otherwise specified. J Child Neurol 22:1198–1203
Parmeggiani A, Barcia G, Posar A, Raimondi E, Santucci M, Scaduto MC (2010) Epilepsy and EEG paroxysmal abnormalities in autism spectrum disorders. Brain Dev 32:783–789
Patry G, Naquet R (1971) Action of dipropylacetic acid upon photosensitive Papio papio. Can J Physiol Pharmacol 49:568–572
Patzold LM, Richdale AL, Tonge BJ (1998) An investigation into sleep characteristics of children with autism and Asperger's Disorder. J Paediatr Child Health 34:528–533
Pineda JA, Juavinett A, Datko M (2012) Self-regulation of brain oscillations as a treatment for aberrant brain connections in children with autism. Med Hypotheses 79:790–798
Polimeni MA, Richdale AL, Francis AJ (2005) A survey of sleep problems in autism, Asperger's disorder and typically developing children. J Intellect Disabil Res 49:260–268
Rapin I (1996) Practitioner review: developmental language disorders: a clinical update. J Child Psychol Psychiatry 37:643–655
Ray T, Tobias JD (2008) Dexmedetomidine for sedation during electroencephalographic analysis in children with autism, pervasive developmental disorders, and seizure disorders. J Clin Anesth 20:364–368
Richdale AL (1999) Sleep problems in autism: prevalence, cause, and intervention. Dev Med Child Neurol 41:60–66
Richdale AL, Schreck KA (2009) Sleep problems in autism spectrum disorders: prevalence, nature, & possible biopsychosocial aetiologies. Sleep Med Rev 13:403–411
Rossi PG, Parmeggiani A, Bach V, Santucci M, Visconti P (1995) EEG features and epilepsy in patients with autism. Brain Dev 17:169–174
Rutter ML (1970) Psycho-social disorders in childhood, and their outcome in adult life. J R Coll Physicians Lond 4:211–218
Ruysschaert L, Warreyn P, Wiersema JR, Oostra A, Roeyers H (2014) Exploring the role of neural mirroring in children with autism spectrum disorder. Autism Res 7(2):197–206. doi:10.1002/aur.1339. Epub 2014 Feb 10
Schreck KA, Mulick JA (2000) Parental report of sleep problems in children with autism. J Autism Dev Disord 30:127–135
Schreck KA, Mulick JA, Smith AF (2004) Sleep problems as possible predictors of intensified symptoms of autism. Res Dev Disabil 25:57–66
Segawa M, Nomura Y (1992) Polysomnography in the Rett syndrome. Brain Dev 14(Suppl):S46–S54
Shimizu Y, Niwa S, Ohta M, Kurita H, Saito Y (1982) Quantitative analysis of awake EEG in infantile autism. Seishin Shinkeigaku Zasshi 84:545–558
Silver WG, Rapin I (2012) Neurobiological basis of autism. Pediatr Clin North Am 59:45–61, x
Smalley SL (1998) Autism and tuberous sclerosis. J Autism Dev Disord 28:407–414
Southgate V, Hamilton AF (2008) Unbroken mirrors: challenging a theory of Autism. Trends Cogn Sci 12:225–229
Spence SJ, Schneider MT (2009) The role of epilepsy and epileptiform EEGs in autism spectrum disorders. Pediatr Res 65:599–606
Sun L, Grutzner C, Bolte S, Wibral M, Tozman T, Schlitt S, Poustka F, Singer W, Freitag CM, Uhlhaas PJ (2012) Impaired gamma-band activity during perceptual organization in adults with autism spectrum disorders: evidence for dysfunctional network activity in frontal-posterior cortices. J Neurosci 32:9563–9573
Szatmari P, Merette C, Emond C, Zwaigenbaum L, Jones MB, Maziade M, Roy MA, Palmour R (2008) Decomposing the autism phenotype into familial dimensions. Am J Med Genet B Neuropsychiatr Genet 147B:3–9

Tanguay PE, Ornitz EM, Forsythe AB, Ritvo ER (1976) Rapid eye movement (REM) activity in normal and autistic children during REM sleep. J Autism Child Schizophr 6:275–288

Tharp BR (2004) Epileptic encephalopathies and their relationship to developmental disorders: do spikes cause autism? Ment Retard Dev Disabil Res Rev 10:132–134

Thirumalai SS, Shubin RA, Robinson R (2002) Rapid eye movement sleep behavior disorder in children with autism. J Child Neurol 17:173–178

Toth K, Stobbe G (2011) Diagnosis of autism spectrum disorders. Pediatr Ann 40:488–492

Trevathan E (2004) Seizures and epilepsy among children with language regression and autistic spectrum disorders. J Child Neurol 19(Suppl 1):S49–S57

Tuchman RF (1997) Language disorders: is EEG clinically useful?. Rev Neurol 25:744–749

Tuchman R (2004) AEDs and psychotropic drugs in children with autism and epilepsy. Ment Retard Dev Disabil Res Rev 10:135–138

Tuchman RF, Rapin I (1997) Regression in pervasive developmental disorders: seizures and epileptiform electroencephalogram correlates. Pediatrics 99:560–566

Tuchman R, Rapin I (2002) Epilepsy in autism. Lancet Neurol 1:352–358

Tuchman R, Cuccaro M, Alessandri M (2010) Autism and epilepsy: historical perspective. Brain Dev 32:709–718

Volkmar FR, McPartland JC (2014) From kanner to DSM-5: autism as an evolving diagnostic concept. Annu Rev Clin Psychol 10:193–212. doi:10.1146/annurev-clinpsy-032813-153710. Epub 2013 Dec 9

Volkmar FR, Nelson DS (1990) Seizure disorders in autism. J Am Acad Child Adolesc Psychiatry 29:127–129

Volkmar F, Chawarska K, Klin A (2005) Autism in infancy and early childhood. Annu Rev Psychol 56:315–336

Wang J, Barstein J, Ethridge LE, Mosconi MW, Takarae Y, Sweeney JA (2013) Resting state EEG abnormalities in autism spectrum disorders. J Neurodev Disord 5:24–25

Wass S (2011) Distortions and disconnections: disrupted brain connectivity in autism. Brain Cogn 75:18–28

Werner E, Dawson G, Munson J, Osterling J (2005) Variation in early developmental course in autism and its relation with behavioral outcome at 3–4 years of age. J Autism Dev Disord 35:337–350

White Pt, Demyer W, Demyer M (1964) Eeg abnormalities in early childhood schizophrenia: a double-blind study of psychiatrically disturbed and normal children during promazine sedation. Am J Psychiatry 120:950–958

Wiggs L, Stores G (2004) Sleep patterns and sleep disorders in children with autistic spectrum disorders: insights using parent report and actigraphy. Dev Med Child Neurol 46:372–380

Chapter 5
Autism Spectrum Disorder. A Clinical Neurophysiology Approach II

Margarita Minou Baez Martin, Lilia María Morales Chacón and Ivette Cabrera Abreu

Abstract One of the most controversial issues in the study of autism spectrum disorder (ASD) is the involvement and dysfunction of sensory systems that underlie atypical behaviors, which could be evaluated through neuroimaging and electrophysiological techniques. Auditory disturbances are the most frequently reported sensory deficits that could explain language and communication deficits shown in ASD patients. Event-related potentials such as auditory brainstem response, mismatch negativity, P50, and P300 have shown differences between patients and normal subjects, supporting the existence of an abnormal auditory processing. Somesthetic perception is also distorted in this heterogeneous group of patients, probably associated with deficits in communication, motor ability, and social skills. It has been confirmed by abnormalities in short- and long-latency somatosensory evoked potentials. Visual processing impairment has also been described in children and young adults with autism. Dipole source analysis revealed that the visual cortex, fusiform gyrus, and medial prefrontal lobe are less active in autism compared with control subjects during the execution of emotion processing tasks. ASD patients also have problems to integrate information from multiple sensory sources necessary to a successful social behavior, and consequently, they show deficits with social and cognitive processes. Electrophysiological and imaging techniques may constitute useful tools in the diagnosis, classification, and therapeutic strategies of ASD patients, considering the diversity of this spectrum.

Keywords Auditory brainstem response · Autism spectrum disorder · Middle latency response · Sensory integration · Somatosensory evoked response · Somesthetic perception · Visual perception

---

M. M. Baez Martin (✉) · L. M. Morales Chacón · I. Cabrera Abreu
International Center for Neurological Restoration (CIREN),
Clinical Neurophysiology Service, Havana, Cuba
e-mail: minou@neuro.ciren.cu

L. M. Morales Chacón
e-mail: lilia.morales@infomed.sld.cu

I. Cabrera Abreu
e-mail: ivettec@neuro.ciren.cu

M. d. l. A. Robinson-Agramonte (ed.), *Translational Approaches to Autism Spectrum Disorder*, DOI 10.1007/978-3-319-16321-5_5

Abbreviations

ABR	Auditory brainstem response
AD/HD	Attention deficit/hyperactivity disorder
ASD	Autism spectrum disorder
MLR	Middle latency response
MRI	Magnetic resonance imaging
PDD	Pervasive development disorder
S-SEPs	Short-latency somatosensory evoked potentials.
TEOAEs	Transient evoked otoacoustic emissions

5.1 Introduction

Autism spectrum disorder (ASD) refers to a group of disorders including autistic disorder, Asperger's disorder, and pervasive developmental disorder (American Psychiatric Association 2013).

One of the most controversial issues in the study of ASD is the involvement and dysfunction of sensory systems that underlie atypical behaviors.

In an extensive review of the literature published in 2008 (1300 reports), Hughes summarizes this contradiction finding equally a low and high frequency of sensory symptoms in patients with autism, and hyper- or hypofunctioning of the sensory systems (Hughes 2009).

Techniques like electroencephalography (EEG), magnetoencephalography (MEG), and functional Magnetic Resonance Imaging (fMRI) have demonstrated their utility to evaluate unimodal sensory processing and multisensory integration (Marco et al. 2011). Neurophysiological responses to auditory, tactile, and visual stimuli in autistic patients have been studied to explain the neural bases of the atypical behaviors previously mentioned (Marco et al. 2011). Other sensory processing deficits, including olfactory and gustatory, have been also described in children with ASD (Tomchek and Dunn 2007; Bennetto et al. 2007).

This chapter aims to summarize the electrophysiological and imaging evidences published in the past years that support the existence of sensory deficits in ASD patients.

5.2 Auditory Perception

Disturbances in auditory processing are the most frequently reported sensory deficits in patients with ASD. In fact, children with autism show a major impairment during the processing of auditory and verbal information compared with visual information (Duncan et al. 2009).

The main trends point out the disturbances of auditory perception of linguistic and social auditory stimulus, and enhanced perception of pitch and music. These discordant findings in auditory abnormality reported in ASD might be the consequence of differences between global and local processing of auditory information occurring in these patients (Kellerman et al. 2005).

Diverse studies in ASD have been designed to test the integrity of the auditory pathway using electrophysiological techniques, such as auditory brainstem response (ABR), but the results are not coincident. In consequence, some authors state that these studies do not support the involvement of brainstem in autism, despite the presence of peripheral hearing impairment in a nonnegligible number of autistic subjects (Klin 1993). For example, Novick et al. show that abnormalities in the brainstem are scarce in patients with ASD using evoked potentials to clicks, pitch changes, and tones. These authors suggest that the registration and storage of stimulus information in the higher levels of processing are more engaged, and they could explain the severe language disorders in ASD (Novick et al. 1980).

Most reports support the hypothesis of impairment in the brainstem auditory pathway to explain the disturbances in auditory processing demonstrated in ASD subjects (Kwon et al. 2007; Magliaro et al. 2010; Maziade et al. 2000; Tas et al. 2007).

The objective evaluation of hearing in a group of children with ASD was carried out by Taset al. (2007) using transient evoked otoacoustic emissions (TEOAEs) and ABR. Both these techniques were obtained in the majority of ASD children bilaterally (83.3 %), while the remaining cases showed an increase in duration of III–V interpeak latency compared with a control group (Tas et al. 2007). Similar results were reported by Kwon et al. (2007) studying 121 autistic children compared with a control group. They reported a prolonged latency of wave V, and an increase in duration of III–V interpeak interval in the ASD group, suggesting a dysfunction or immaturity of the central auditory pathway, particularly at the level of mesencephalon (Kwon et al. 2007).

Autistic patients and unaffected first degree relatives showed the prolongation of I–III interpeak interval of ABR in relation to control subjects. So, this finding could be proposed as an endophenotype of the disease (Maziade et al. 2000).

The auditory abnormalities mentioned could contribute to sensory deprivation and consequently to communication deficits. Furthermore, aberrations in the perception and processing of the audiological stimuli, like maturational defects and atypical lateralization, has been proposed (Hitoglou et al. 2010).

Other techniques have been used to assess hearing in autistic subjects: pure tone and speech audiometry, acoustic immitance measures, ABR, auditory middle latency response (MLR), and cognitive potential (P300). All auditory evoked potentials showed differences with respect to a group of normal subjects. Particularly, statistically significant differences between groups were observed in the latency of waves III and V, and in the duration of interpeak intervals I–III and I–V of ABR, confirming the impairment of the brainstem auditory pathway. The absence of P300 component suggests the dysfunction of cortical areas in the same patients (Magliaro et al. 2010).

Abnormal auditory symptoms include peripheral and sensorineural hearing deficit, both hypo- and hypersensitivity to auditory stimulation with bizarre reactions to sounds, and limitations to filter relevant auditory information in the presence of environmental background noise (Källstrand et al. 2010). Khalfa et al. explained that difficulty in filtering information presupposes the participation of active cochlear mechanisms regulated by the efferent olivocochlear system and evaluated with TEOAEs. The abnormalities in lateralization at the auditory periphery observed in ASD patients are probably related to dysfunction at higher levels of the auditory processing mediated by the medial olivocochlear system (Khalfa et al. 2001).

Limitations to filter auditory information in a noisy background also include speech. Evidence of this was provided by Russo et al., evaluating speech-evoked responses in quiet and noise background in ASD patients and typically developing children. Differences between quiet and noise responses were minimal in ASD children, probably related to a severe depression of the responses in a noise background (Russo et al. 2009).

On the contrary, other authors reported that nonmentally retarded children with autism had a sensory gating to auditory stimulus similar to a control group using the P50 gating paradigm (Kemner et al. 2002). All these apparent differences in published papers are probably related to the nature of the auditory stimuli used.

Another aspect that has to be taken into account is the automatic and active processing of auditory stimuli evaluated with the mismatch negativity. The amplitude of this event-related potential is smaller in autistic than in normal children in unattended conditions. On the other hand, when the stimuli are attended the differences between groups disappear. All these results confirmed the existence of an abnormal automatic auditory processing in autism (Dunn et al. 2008).

Larger P300 component to target auditory stimuli has been also found in autistic children during active conditions with respect to healthy, dyslexic, and attention deficit/hyperactivity disorder (AD/HD) groups, while other authors found larger P300a in younger children with respect to older (Hughes 2009).

An interesting finding reported by Lai et al. in low-functional autistic patients showed that functional systems which process speech and song are more related to the last one. Some structures, like the left inferior frontal gyrus, are more activated during song stimulation and less with speech stimulation in autistic patients compared with control normal subjects. Measures like functional connectivity between this gyrus and superior temporal gyrus was increased for songs compared to speech in patients. At the same time, there were increased frontal-posterior connections. These and other findings of this study support the assumption that functional systems related to speech and song are more active for song in this subgroup of autistic patients (Lai et al. 2012).

Deficits in communication in ASD patients include a variety of disturbances in language, ranging from semantic-pragmatic deficits to the absence of speech. But these problems are probably dependent, at least in part, on auditory aberrant perception (Kujala et al. 2013).

Deficient prosody, for example, is a distinguishing characteristic of the language impairment in ASD patients related to acoustic cues such as pitch contour. The sensory encoding of pitch was evaluated by evoked brainstem responses to pitch syllables in a group of ASD patients with normal intelligence and hearing, some of them showed deficient pitch tracking compared with normal children. Obviously, there is a subcortical origin of prosody encoding deficits demonstrated by electrophysiological techniques. This should be taken into account in the diagnosis and treatment of this subgroup of ASD patients (Russo et al. 2008).

Of course, ASD include a wide spectrum of entities with a diversity of responses. Matas et al. studied the audiological and electrophysiological profile of patients with autism and Asperger syndrome, and compared their results with a group of normal subjects. They did not find abnormalities in the audiological evaluation in all subjects, but 50% of patients with autism and 30% with Asperger syndrome showed ABR alterations, with statistically significant differences between groups. Auditory MLR presented abnormalities in all groups without differences between them, while P300 component showed differences between Asperger patients and control subjects. These results suggest that auditory information processing at different levels of the pathway has specific profiles in the diverse forms of the disease (Matas et al. 2009).

Other authors have found normal hearing levels in high-functioning autistic children with a pervasive development disorder (PDD) and speech delay (Psillas et al. 2006).

Bruneau et al. demonstrated in a group of mentally retarded children with ASD the existence of electroclinical correlations between the amplitude of the right temporal N1 cortical response and the verbal and nonverbal communication abilities, suggesting a reorganization of the hemisphere functions, with a greater activation of the right hemisphere for functions that correspond to the left one. This especially was true for the lateral aspect of the superior temporal gyrus which belongs to secondary auditory areas (Bruneau et al. 2003).

5.3 Somesthetic Perception

In comparison with auditory disturbances in ASD patients, fewer studies have been conducted to evaluate the dysfunction of somesthetic perception. Some of the symptoms include hyper- or hypoesthesia to touch, pain, and temperature (Miyazaki et al. 2007). Somatosensory input is fundamental for motor development, while touch is also critical for healthy social and communication skills in early childhood and beyond.

Short-latency somatosensory evoked potentials (S-SEPs) elicited by the stimulation of median nerve were evaluated and correlated with clinical symptoms in a group of 24 children with autism. The abnormalities include prolonged peak latency of N20, increase in central conduction time, and a prevalence of the amplitude of N20/P25 in the right hemisphere (more than twofold) that indicates hyperactivity in the right primary somatosensory area of these patients (Miyazaki et al. 2007).

Similar results have been published by Cabrera et al., who obtained S-SEPs with stimulation of posterior tibial nerve. In this study, 92 % of ASD children showed abnormalities related to an increase of the central conduction time (65 %) and morphological distortion or absence of the evoked cortical response (Cabrera et al. 2011).

It has also been reported a strong positive valence to tactile stimuli, which provokes the excitation of patients, and enhanced responsiveness mediated by hypersensitivity to vibration and thermal pain (Hughes 2008).

In 2010, Cascio published a review about somatosensory processing in neurodevelopmental disorders that include autism. This author found that somatosensory perception is aberrant in ASD, associated with deficits in communication, motor ability, and social skills, which should be considered in the etiology and treatment of these disorders (Cascio 2010).

More surprising were the results reported by Wong et al., who found slower and larger amplitude of the event-related potential P300 in the parietal somatosensory cortices of children with autism during the processing of facial gender and emotion, besides the activation of visual cortex and fusiform gyrus (Wong et al. 2008). These results suggest an abnormal specialization of the cortex engaged to social brain networks.

5.4 Visual Perception

Visual processing abnormalities have also been described in children and young adults with autism. For example, Milne et al. stated that perception of simple visual stimuli is atypical in ASD patients (Milne et al. 2009). These authors found differences between patients and control subjects in the electroencephalographic activity during visual perception with a cortical topographic differentiation (striate-extrastriate cortex and cingulate gyrus) (Milne et al. 2009).

There are specific abnormalities in early processing of visual stimuli that are probably related to perception. It has been demonstrated in patients with PDD, verifying a reduction in amplitude of event-related potentials at the occipital cortex in response to visual stimuli with high and low spatial frequencies (Kemner and van Engeland 2006).

More specific results were described by Boeschoten et al. using short-latency visual-evoked potentials and dipole sources of their components. They found abnormalities for high spatial frequency stimuli during the early processing, and in late phases for both high and low spatial frequency stimuli (Boeschoten et al. 2007a). This atypical response might suggest the existence of difficulties in visual processing from the primary levels of analysis in the visual pathway of PDD subjects.

The same authors explained that the previously described anomalies had an effect in the abnormal face processing in PDD subjects, provoking disturbances of social behavior (Boeschoten et al. 2007b).

Nevertheless, most papers support the idea of a higher level of dysfunction that involves extrastriate cortices (Deeley et al. 2007; Gunji et al. 2009; Koshino et al. 2008; Pierce and Redcay 2008; Wong et al. 2008; O'Connor et al. 2007).

Event-related functional MRI demonstrated a lesser activation of fusiform and extrastriate cortices to detect facial expressions of primary emotions in Asperger syndrome compared with control subjects (Deeley et al. 2007).

The distinction between self, familiar, and unfamiliar faces is disturbed in PDD patients as shown in the study of Gunji et al., whose results suggest a deficit of semantic encoding of faces in this group of patients (Gunji et al. 2009).

When recognition of faces has an emotional connotation the results differ. Event-related potentials that evaluate recognition of emotional facial expressions (N1,N170, and P2 components) are in general not affected in ASD, but dipole source analysis revealed that the visual cortex, fusiform gyrus, and medial prefrontal lobe are less active in autism compared with control subjects during the execution of emotion processing tasks (Wong et al. 2008).

Autistic patients need to make an excessive processing of the visual information to successfully differentiate target and a novel stimuli during the execution of a three-stimulus oddball paradigm task (Sokhadze et al. 2009).

5.5 Integration of Sensory Information

The impairment in social cognition and the abnormalities in sensory and perceptual processing are characteristics of patients with ASD. These subjects especially have problems to integrate information from multiple sensory sources necessary to a successful social behavior, and consequently, they show deficits with social and cognitive processes.

ASD patients can also have problems with earlier stages of multisensory information processing, evidenced by the recording of the cerebral electrical activity and the reaction time during a simple audiovisual task (Brandwein et al. 2013).

It has been explained that communicative impairment of PDD subjects is based on a disturbed complex audiovisual integration of speech stimuli, while low-level abilities integration is relatively spared (Magnee et al. 2008).

Auditory-somatosensory integration is also damaged in ASD patients compared to matched typically developing children, different to auditory or somatosensory stimuli alone (Russo et al. 2010).

Visual-somatosensory integration was evaluated by Kemner et al. using a three-stimulus oddball task (standard, deviant, and novel stimuli) in autistic and control children. They reported that autistic children differed from controls with respect to visual (P2, N2, and P3) and somatosensory (P3) responses, showing abnormal processing of proximal and distal stimuli, but they could not find evidence of abnormal lateralization of the event-related potential components (Kemner et al. 1994).

5.6 Conclusions

The main sensory modalities and multisensory processing are impaired in autistic patients, conducting to a global sensory dysfunction that could be in the background of other disturbances such as language and social disabilities.

Electrophysiological and imaging techniques may constitute useful tools in the diagnosis, classification, and therapeutic strategies of ASD patients, considering the diversity of this spectrum.

Acknowledgments We thank Odalys Morales-Chacon for her English assistance and Jorge Bergado-Rosado for his contributions.

References

American Psychiatric Association (2013) American psychiatric association: diagnostic and statistical manual of mental disorders, DSM-5. American Psychiatric Publishing, Washington, DC

Bennetto L, Kuschner ES, Hyman SL (2007) Olfaction and taste processing in autism. Biol Psychiatry 62:1015–1021

Boeschoten MA, Kenemans JL, van Engeland H, Kemner C (2007a) Abnormal spatial frequency processing in high-functioning children with pervasive developmental disorder (PDD). Clin Neurophysiol 118:2076–2088

Boeschoten MA, Kenemans JL, van Engeland H, Kemner C (2007b) Face processing in pervasive developmental disorder (PDD): the roles of expertise and spatial frequency. J Neural Transm 114:1619–1629

Brandwein AB, Foxe JJ, Butler JS, Russo NN, Altschuler TS, Gomes H, Molholm S (2013) The development of multisensory integration in high-functioning autism: high-density electrical mapping and psychophysical measures reveal impairments in the processing of audiovisual inputs. Cereb Cortex 23:1329–1341

Bruneau N, Bonnet-Brilhault F, Gomot M, Adrien JL, Barthelemy C (2003) Cortical auditory processing and communication in children with autism: electrophysiological/behavioral relations. Int J Psychophysiol 51:17–25

Cabrera I, Báez M, Maragoto C, Galvizu R, Vera H, Ortega M (2011) Evaluación funcional de sistemas sensoriales mediante potenciales evocados en niños con trastornos del espectro autista. Enfermería Global 24:39–44

Cascio CJ (2010) Somatosensory processing in neurodevelopmental disorders. J Neurodevelop Disord 2:62–69

Deeley Q, Daly EM, Surguladze S, Page L, Toal F, Robertson D, Curran S, Giampietro V, Seal M, Brammer MJ, Andrew C, Murphy K, Phillips ML, Murphy DG (2007) An event related functional magnetic resonance imaging study of facial emotion processing in Asperger syndrome. Biol Psychiatry 62:207–217

Duncan CC, Barry RJ, Connolly JF, Fischer C, Michie PT, Naatanen R, Polich J, Reinvang I, Van Petten C (2009) Event-related potentials in clinical research: guidelines for eliciting, recording, and quantifying mismatch negativity, P300, and N400. Clin Neurophysiol 120:1908

Dunn MA, Gomes H, Gravel J (2008) Mismatch negativity in children with autism and typical development. J Autism Dev Disord 38:52–71

Gunji A, Inagaki M, Inoue Y, Takeshima Y, Kaga M (2009) Event-related potentials of self-face recognition in children with pervasive developmental disorders. Brain Dev 31:139–147

Hitoglou M, Ververi A, Antoniadis A, Zafeiriou DI (2010) Childhood autism and auditory system abnormalities. Pediatr Neurol 42:309–314

Hughes JR (2008) A review of recent reports on autism: 1000 studies published in 2007. Epilepsy Behav 13:425–437
Hughes JR (2009) Update on autism: a review of 1300 reports published in 2008. Epilepsy Behav 16:569–589
Källstrand J, Olsson O, Fristedt Nehlstedt S, Ling Sköld M, Nielzén S (2010) Abnormal auditory forward masking pattern in the brainstem response of individuals with Asperger syndrome. Neuropsychiatr Dis Treat 6:289–296
Kellerman GR, Fan J, Gorman JM (2005) Auditory abnormalities in autism: toward functional distinctions among findings. CNS Spectr 10:748–756
Kemner C, van Engeland H (2006) ERPs and eye movements reflect atypical visual perception in pervasive developmental disorder. J Autism Dev Disord 36:45–54
Kemner C, Verbaten MN, Cuperus JM, Camfferman G, van Engeland H (1994) Visual and somatosensory event-related brain potentials in autistic children and three different control groups. Electroencephalogr Clin Neurophysiol 92:225–237
Kemner C, Oranje B, Verbaten MN, van Engeland H (2002) Normal P50 gating in children with autism. J Clin Psychiatry 63:214–217
Khalfa S, Bruneau N, Roge B, Georgieff N, Veuillet E, Adrien JL, Barthelemy C, Collet L (2001) Peripheral auditory asymmetry in infantile autism. Eur J Neurosci 13:628–632
Klin A (1993) Auditory brainstem responses in autism: brainstem dysfunction or peripheral hearing loss? J Autism Dev Disord 23:15–35
Koshino H, Kana RK, Keller TA, Cherkassky VL, Minshew NJ, Just MA (2008) fMRI investigation of working memory for faces in autism: visual coding and underconnectivity with frontal areas. Cereb Cortex 18:289–300
Kujala T, Lepisto T, Naatanen R (2013) The neural basis of aberrant speech and audition in autism spectrum disorders. Neurosci Biobehav Rev 37:697–704
Kwon S, Kim J, Choe BH, Ko C, Park S (2007) Electrophysiologic assessment of central auditory processing by auditory brainstem responses in children with autism spectrum disorders. J Korean Med Sci 22:656–659
Lai G, Pantazatos SP, Schneider H, Hirsch J (2012) Neural systems for speech and song in autism. Brain 135:961–975
Magliaro FCL, Scheuer CI, Assumpção Júnior FB, Matas CG (2010) Study of auditory evoked potentials in autism (original title: Estudo dos potenciais evocados auditivosem autismo). Pró-Fono Revista de Atualização Científica 22:31–36
Magnee MJ, de Gelder B, van Engeland H, Kemner C (2008) Audiovisual speech integration in pervasive developmental disorder: evidence from event-related potentials. J Child Psychol Psychiatry 49:995–1000
Marco EJ, Hinkley LB, Hill SS, Nagarajan SS 2011 Sensory processing in autism: a review of neurophysiologic findings. Pediatr Res 69:48R–54R
Matas CG, Goncalves IC, Magliaro FC (2009) Audiologic and electrophysiologic evaluation in children with psychiatric disorders. Braz J Otorhinolaryngol 75:130–138
Maziade M, Merette C, Cayer M, Roy MA, Szatmari P, Cote R, Thivierge J (2000) Prolongation of brainstem auditory-evoked responses in autistic probands and their unaffected relatives. Arch Gen Psychiatry 57:1077–1083
Milne E, Scope A, Pascalis O, Buckley D, Makeig S (2009) Independent component analysis reveals atypical electroencephalographic activity during visual perception in individuals with autism. Biol Psychiatry 65:22–30
Miyazaki M, Fujii E, Saijo T, Mori K, Hashimoto T, Kagami S, Kuroda Y (2007) Short-latency somatosensory evoked potentials in infantile autism: evidence of hyperactivity in the right primary somatosensory area. Dev Med Child Neurol 49:13–17
Novick B, Vaughan HG Jr, Kurtzberg D, Simson R (1980) An electrophysiologic indication of auditory processing defects in autism. Psychiatry Res 3:107–114
O'Connor K, Hamm JP, Kirk IJ (2007) Neurophysiological responses to face, facial regions and objects in adults with Asperger's syndrome: an ERP investigation. Int J Psychophysiol 63:283–293

Pierce K, Redcay E (2008) Fusiform function in children with an autism spectrum disorder is a matter of "who". Biol Psychiatry 64:552–560
Psillas G, Psifidis A, Antoniadou-Hitoglou M, Kouloulas A (2006) Hearing assessment in pre-school children with speech delay. Auris Nasus Larynx 33:259–263
Russo NM, Skoe E, Trommer B, Nicol T, Zecker S, Bradlow A, Kraus N (2008) Deficient brainstem encoding of pitch in children with autism spectrum disorders. Clin Neurophysiol 119:1720–1731
Russo N, Zecker S, Trommer B, Chen J, Kraus N (2009) Effects of background noise on cortical encoding of speech in autism spectrum disorders. J Autism Dev Disord 39:1185–1196
Russo N, Foxe JJ, Brandwein AB, Altschuler T, Gomes H, Molholm S (2010) Multisensory processing in children with autism: high-density electrical mapping of auditory-somatosensory integration. Autism Res 3:253–267
Sokhadze E, Baruth J, Tasman A, Sears L, Mathai G, El-Baz A, Casanova MF (2009) Event-related potential study of novelty processing abnormalities in autism. Appl Psychophysiol Biofeedback 34:37–51
Tas A, Yagiz R, Tas M, Esme M, Uzun C, Karasalihoglu AR (2007) Evaluation of hearing in children with autism by using TEOAE and ABR. Autism 11:73–79
Tomchek S, Dunn W (2007) Sensory processing in children with and without autism: a comparative study using the short sensory profile. Am J Occup Ther 61:190–200
Wong TK, Fung PC, Chua SE, McAlonan GM (2008) Abnormal spatiotemporal processing of emotional facial expressions in childhood autism: dipole source analysis of event-related potentials. Eur J Neurosci 28:407–416

Chapter 6
Neuroimages in Autism

Rafael Rodríguez-Rojas, Calixto Machado, Karla Batista, Maylen Carballo and Gerry Leisman

Abstract This chapter summarizes findings of the morphological and functional brain changes in autistic spectrum disorders (ASD), demonstrated by a growing body of researches that employs diverse neuroimaging methods. We discuss major anatomical findings from studies in ASD utilizing structural magnetic resonance imaging (MRI) and diffusion tensor imaging (DTI). We also explore findings from functional MRI (fMRI) studies, including task-related activation studies and functional connectivity. Regional changes in brain metabolism are examined using magnetic resonance spectroscopy (MRS) and nuclear medicine techniques, including positron emission tomography (PET) and single-photon emission tomography (SPECT). Finally, we highlight directions for continued research in neuroimaging of ASD at a network level using graph theoretical approaches.

Keywords Magnetic resonance imaging · Diffusion tensor imaging · Spectroscopy · Emission tomography · Connectivity · ASD

6.1 Introduction

During the last decades, neuroimaging studies have improved our knowledge of brain development and contributed to our understanding of disorders involving the developing brain like autism spectrum disorder (ASD). Morphological studies have provided evidence of structural differences in ASD compared with the normal population. In parallel, functional neuroimaging methods attempt to elucidate the neurobiology and neurochemistry of autistic disorders.

R. Rodríguez-Rojas (✉) · K. Batista · M. Carballo
Brain Images Processing Group, International Center for Neurological Restoration, Havana, Cuba
e-mail: rafael.rguez@infomed.sld.cu

C. Machado
Neurophysiology Department, Institute of Neurology and Neurosurgery, Havana, Cuba

G. Leisman
The National Institute for Brain and Rehabilitation Sciences, Nazareth, Israel

O.R.T.-Bruade College of Engineering, Karmiel, Israel

M. d. l. A. Robinson-Agramonte (ed.), *Translational Approaches to Autism Spectrum Disorder,* DOI 10.1007/978-3-319-16321-5_6

The goal of this chapter is to provide an overview of the morphological and functional brain changes in ASD, demonstrated by a growing body of researches that employ diverse neuroimaging methods. We discuss major anatomical findings from studies in ASD utilizing structural magnetic resonance imaging (MRI) and diffusion tensor imaging (DTI). We also explore findings from functional MRI (fMRI) studies, including task-related activation studies and functional connectivity. Regional changes in brain metabolism are examined using magnetic resonance spectroscopy (MRS) and nuclear medicine techniques, including positron emission tomography (PET) and single-photon emission tomography (SPECT). Finally, we highlight directions for continued research in neuroimaging of ASD at a network level using graph theoretical approaches.

6.2 Structural Magnetic Resonance Imaging

Image-based morphometric analysis with statistical parametric mapping techniques permits rapid, largely automated, whole-brain analysis of autistic neuroanatomy. There are converging evidences that young children with ASD have a significant increase in cerebral volume in both gray and white matter. However, the results from studies of adolescents and adults with ASD are inconsistent. It is possible that discrepant findings primarily due to the differences in patient's cohorts within the autistic spectrum who exhibit different patterns of structural abnormalities, even though clinical symptoms are similar. Alternatively, this difference might also arise from different techniques of image processing and analysis. Based on the analysis methods applied to the image data, these studies can be classified into region of interest (ROI)-based, voxel (VBM)-based, surface (SBM)-based, or tensor (TBM)-based morphometry. In all these approaches, segmentation of brain structures is an essential step for morphometric analysis.

6.2.1 ROI-Based Morphometry

A variety of methods have been used for segmenting regions of interest from MRI brain studies in ASD populations (ASDs). Common procedures include manual drawing by experts, atlas-based segmentation or volumes quantification of tissue types (e.g., gray and white matter or total brain volume). For an extensive review, see Ref. (Chen et al. 2011; Brambilla et al. 2003). Increased total brain, parieto-temporal lobe, and cerebellar hemisphere volumes have been the most replicated abnormalities in autism. However, whether or not this abnormal enlargement persists into later childhood and adolescence is still a matter of debate. Enlargement of amygdala in children, and reduced volumes in subregions of the corpus callosum and hippocampal area have also been reported in the autistic individuals (Chen et al. 2011; Brambilla et al. 2003).

6.2.2 Voxel-Based and Tensor-Based Morphometry

Voxel-based morphometry (VBM) computes the probability of each voxel being associated with a specific tissue type. Maps of each tissue are smoothed, and statistical tests are performed at each voxel representative of a specific brain region, to determine the morphometric changes related to autistic features. Posterior probability is usually called *tissue density*. Tensor-based morphometry (TBM) is a more computationally intensive method, in which a deformation field is obtained for each subject, warping their brain to match a common brain template, and encoding the relative positions of various brain landmarks.

Several VBM and TBM studies have been performed in ASD, reporting white and gray matter abnormalities in a widely distributed network, including the frontal, temporal, parietal and occipital lobes, cingulate gyrus, fusiform gyrus, hippocampus and parahippocampal regions, cerebellum, internal capsule, caudate nuclei, and corpus callosum (see (Nickl-Jockschat et al. 2012; Cauda et al. 2014; Dennis and Thompson 2013) for review). However, morphometric findings are inhomogeneous and even contradictory. Early studies showed, decreased gray matter density in the left medial temporal lobe (Brieber et al. 2007), right inferior temporal gyrus, entorhinal cortex, and rostral fusiform gyrus in adolescent (Kwon et al. 2004), and as higher grey matter volumes in the left inferior parietal cortex and the right supramarginal gyrus in children and adolescents fulfilling the diagnostic criteria for ASD (Brieber et al. 2007).

Rojas et al. reported the existence of regions exhibiting larger volumes in ASD's in medial frontal gyri, left precentral and right postcentral gyrus, right fusiform gyrus, caudate nuclei, and the left hippocampus, and smaller volumes exclusively in the cerebellum, comparing autistic young adults with age-matched controls (Rojas et al. 2006). On the other hand, Brun et al. (2009) found significantly decreased cortical volumes in the parietal, left temporal, and left occipital lobes, bilaterally, using TBM. Alterations of the basal ganglia, especially the caudate nuclei, have been repeatedly described in ASD patients and were often found to correlate with impaired motor performance or repetitive and stereotyped behavior (for review see (Nickl-Jockschat et al. 2012)).

In contrast to the rather small sample sizes of the previously referenced studies, Nickl-Jockschat et al. included data from, 277 ASD patients and 303 healthy controls in a wide-ranging meta-analysis, to provide a quantitative summary of brain structure changes using VBM (Nickl-Jockschat et al. 2012). They located six significant clusters of convergence indicating disturbances in brain structures of ASD patients, including the lateral occipital lobe, the pericentral region, the medial temporal lobe, the basal ganglia, and proximate to the right parietal operculum. A more recent meta-analysis found significant topological relationships between gray and white matter abnormalities in ASD (Cauda et al. 2014). They localized numerous bilateral negative concordances, but with a higher prevalence in the right hemisphere, while positive concordances were found in the left hemisphere. The emergence of different hemispheric contribution suggests a relation with pathogenetic factors affecting the right hemisphere during early developmental stages. Besides, the authors found an essentially negative concordance and discordance between brains areas involved in social cognition with the white matter fiber tracts linking them. Less white matter concentration in the region of the genu, rostrum, and splenium have been reported in adolescent and young high functioning autistic subjects (Chung et al. 2004).

While inconclusive, VBM and TBM studies have revealed a complex, region-specific and age-related changes of regional volumes in ASD, comprising cortical and subcortical regions and as areas of white matter. These data provide evidence of altered brain morphology in ASD and possibly its role in the pervasive symptoms dramatically impairing communication and social skills in ASD patients.

6.2.3 Surface-Based Morphometry

Surface-based morphometry (SBM) assesses local folding giving access to topological properties of the cortex such as cortical thickness, regional surface area, and sulcal depths. These descriptors reveal different attributes of the complex interactions between brain regions in ASD. The separate analysis of these orthogonal characteristics and the anatomical reliability of surface-based metrics enhance sensitivity and predictive value in comparison with VBM.

Increased cortical thickness has been reported over the entire cerebral cortex in children with autism, primarily driven by increases in parietal and temporal cortices (Hardan et al. 2006). Recent approaches using more fractionated subcomponents of cortical and subcortical anatomy have revealed cortical thickening in regions known to be crucial for social cognition, behavioral regulation, language, and social information, although regional thinning with lesser extent have also been described (Jiao et al. 2010; Hyde et al. 2010; Raznahan et al. 2013). Figure 6.1 summarizes the brain areas with abnormal cortical thickness in children and adolescent with ASD.

These findings suggest that abnormal patterns of cortical thickness in a broader network of cortical areas subserving social cognition, communication, and repetitive behaviors, could be underlying the core features of autism. However, this effect appears to be dynamically age-dependent, as the most pronounced period of cortical enlargement occurs in early postnatal periods (Hazlett et al. 2011; Courchesne et al. 2001, Schumann et al. 2010), while adults with ASD tend to have increased age-related cortical thinning in comparison with typically developing individuals (TD) (Hardan et al. 2009). Conversely, Hazlett et al. found an increase in the surface area temporal, frontal, and parietal-occipital regions in 2 year old children with autism compared to controls, but no significant differences in the cortical thickness were reported (Hazlett et al. 2011).

Gyrification patterns have also been studied in autism. Hardan et al. reported an increased left frontal gyrification index in autistic children and adolescents but not in adults (Hardan et al. 2004). Nordahl et al. explored cortical folding across a range of patients between 7.5 and 18 years (Nordahl et al. 2007). They identified cortical shape abnormalities across all ages, more pronounced in children. The patterns of

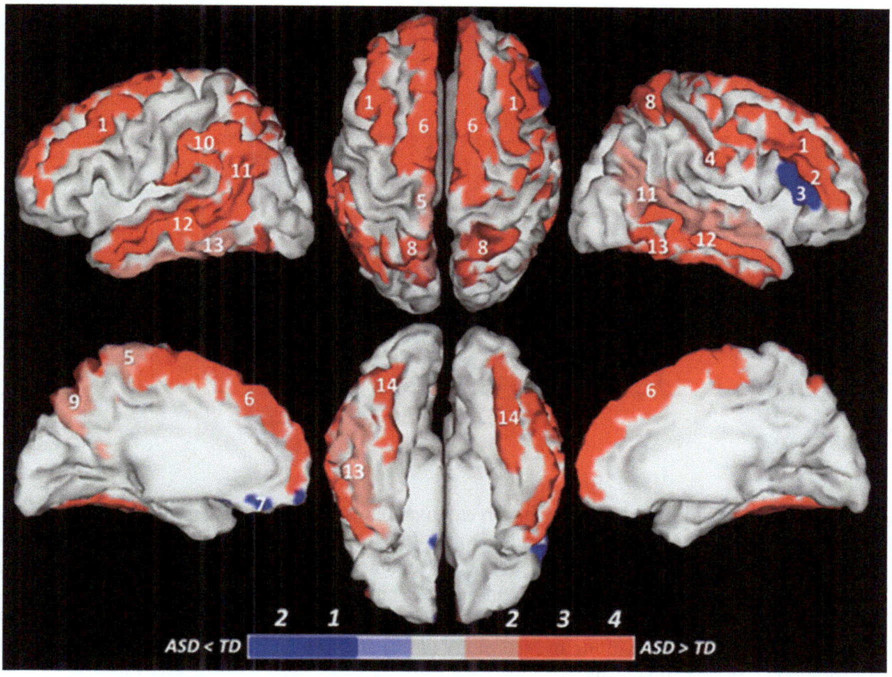

Fig. 6.1 Synthesis of findings from existing studies of cortical thickness differences between children and adolescents with *ASD* and *TD*, integrated using a sulco-gyral-based parcellation. The *red* and *blue* ROIs represent regions for which cortex was thicker and thinner, respectively, in the *ASD* group. *Color scale* represents the number of times each region has been reported in reviewed literature. *1 middle* frontal gyrus, *2* inferior frontal sulcus, *3* triangular part of the inferior frontal gyrus, *4* precentralgyrus, *5* paracentral sulcus, *6* superior frontal gyrus, *7* orbitofrontal gyrus, *8* superior parietal lobule, *9* precuneus, *10* supramarginalgyrus, *11* superior temporal sulcus, *12 middle* temporal gyrus, *13* inferior temporal gyrus, and *14* fusiform gyrus

abnormalities varied between low-functioning autism, high-functioning autism, and Asperger's syndrome.

All of the aforementioned findings taken together, the increased amount of surface-based evidences suggest that early overgrowth in cortical volume in ASD may be associated with a disproportionate enlargement in cortical thickness and surface area across multiple brain regions, with a tendency to normalization at later stages of development. While still inconsistent, these results suggest that disruptions of cortical topology in preschool aged ASD children may underlie a distinct set of pathogenic mechanisms.

6.2.4 Automatic 3D Model-Based Morphometry

An alternative approach combines the characteristics of ROI-based analysis with voxel or surface-based techniques for obtaining the segmentation of individual MRI

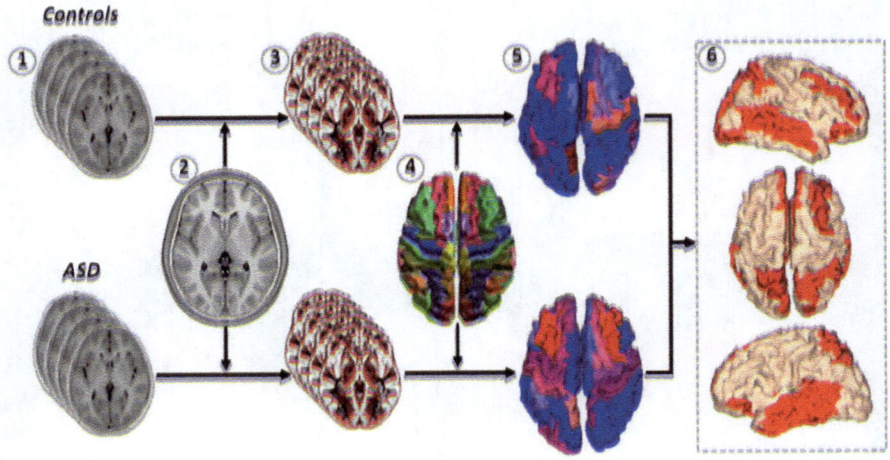

Fig. 6.2 Pipeline and results of an atlas-driven voxel based morphometry. *1* High resolution anatomical MRI. *2* Normalization to a standard space (e.g., *MNI* space). *3* Segmentation and extraction of the gray matter. *4* Atlas-based labeling. *5* Regional mean volumes were calculated via the atlas-based procedure described in (Aleman-Gómez et al. 2006). *6* Dorsal and lateral views of statistical parametric maps showing increased volumes (*red* areas) in *ASD* children. Preliminary *MRI* data were obtained at the Institute of Neurology (*INN*) and International Center for Neurological Restoration (*CIREN*), in Havana

volume into different anatomical structures using a standardized atlas. Structural MRI with high spatial resolution is normalized into a standard space defined by a registration template. The resulting transformation is reversed to map the atlas structures onto the original brain geometry, thus enabling morphometric quantification. Different strategies have been utilized to merge information coming from registration algorithms and tissue classification for obtaining suitable gross cortical structure segmentation. Our group has used an approach based on individual brain atlases using statistical parametric mapping (IBASPM) to address brain morphometry in ASD. Figure 6.2 outlines this procedure. Panel 6 shows preliminary results from our group, consistent with a widespread increase in cortical volume, involving frontal, parietal, and occipitotemporal regions.

6.3 Functional Magnetic Resonance Imaging

The advents of higher magnetic fields and faster acquisition sequences have allowed the acquisition of dynamic data that brings together maximal signal-to-noise ratio and good spatial contrast in conjunction with whole-brain coverage. Functional MRI (fMRI), as opposed to traditional MRI procedures used to study brain anatomy, provides insights into dynamic brain changes having a time course closer to that of neurophysiological activities. The technique therefore has huge potential for addressing novel research questions regarding the abnormalities in human brain function underlying the core symptoms of ASD.

The relatively low temporal resolution and long acquisition time of fMRI challenged the extensive application of this technique in certain groups of patients, like children and low functioning autistic individuals. Despite these concerns, the contribution of fMRI to understanding aspects of the biology of ASD has been invaluable, as it has helped to establish that ASD is a distributed disorder comprising several neural systems. It is beyond the scope of this chapter to detail the extensive literature that explores the patterns of activation in response to dissimilar tasks. We will outline the main approaches and promising developments in the use of fMRI to study functional networks in ASD.

6.3.1 Activation Studies

There is a link between increased local neuronal activity and increased regional cerebral blood flow, blood volume, and blood oxygen content (Heeger and Rees 2002). A slight but measurable local lengthened MR signal is driven by the imbalance between oxygenated and nonoxygenated hemoglobin. This blood-oxygen-level-dependent (BOLD) effect forms the basis for the use of fMRI, allowing the identification of brain regions used during the performance of a specific task. A wide and increasing number of researches have explored the patterns of activation related to core symptoms in ASD, using activation paradigms related to the sound and face processing, cognitive control, theory of mind tasks, language processing, or tasks involving imagery content, among others. Extensive reviews can be seen in the published literature (Anagnostou and Taylor 2011; Dichter 2012; Hugdahl et al. 2012; Stigler et al. 2011; Williams and Minshew 2007; DiMartino et al. 2009). Figure 6.3 shows models of abnormal activity patterns in ASD in response to different paradigms used in fMRI. Note that these models consider the spatial location of the areas where abnormalities in activation have been consistently reported, regardless of the direction of the differences (hypoactivation or hyperactivation) given the variability in results and the limited space.

The fMRI studies in ASDs have provided behavioral evidence of incapacity for emotional judgments about faces (e.g., the inability for reading of the intentions or emotions facially expressed by others) and impaired face recognition and discrimination (e.g., to discriminate between familiar and unfamiliar faces) in ASDs (Dichter 2012; Stigler et al. 2011). ASD-related hypoactivation of the fusiform gyrus (Brodman area BA 37), labeled as the 'fusiform face area', has been reported across both studies of facial form and facial expression perception (Stigler et al. 2011). The response of the amygdala to faces in ASDs has also been studied, with mixed and even inconsistent results (Dichter 2012; Hugdahl et al. 2012). Individuals with ASD have been shown to have difficulty with the ability to infer feeling states and/or intentions, skills that are essentials for an appropriate social interaction. The fMRI studies have revealed that children and adults with autism show reduced activation

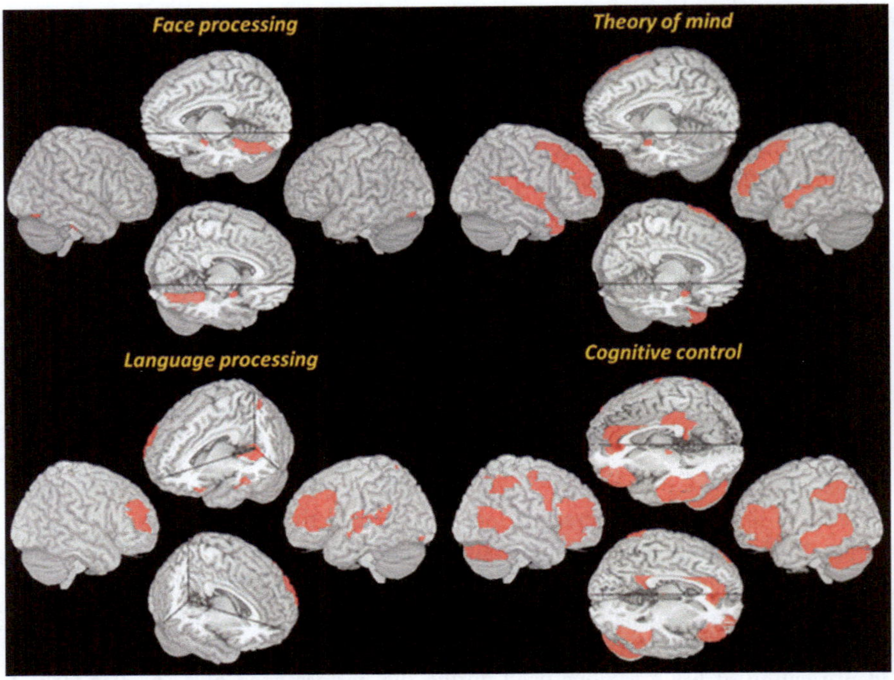

Fig. 6.3 The figure shows brain network activations across studies and experimental manipulations in *ASD* compared to *TD* individuals, visualized on a 3D *MRI* template. Abnormal activation maps were modeled using an MRI-based atlas, showing regions reported to be involved in specific tasks. See text for further explanations

throughout the medial prefrontal cortex, superior temporal sulcus, right temporal pole, and amygdala, engagedin the so-called "theory of mind" network (DiMartino et al. 2009). Amygdala has received particular scrutiny in fMRI studies of theory of mind. Impaired modulation of activity in this area has been associated with the reduced social and communication abilities in ASD.

Autistic spectrum disorders (ASD) are associated with a wide range of language and communication impairments, ranging from a lack of functional speech to outstanding language abilities (Williams and Minshew 2007). fMRI studies of communication deficits in ASDs have focused predominantly on brain regions mediating language perception, comprehension, and generation. Activation paradigms include auditory processing, sentence and language comprehension, verbal fluency, visual imagery, among others. Overall, these data support differential lateralization of language processing and production regions during communication tasks, neurofunctional deficits for speech and recruitment of brain regions that do not typically process language (Anagnostou and Taylor 2011; Dichter 2012; Williams and Minshew 2007). Abnormal activation in the left inferior frontal gyrus (including Broca's area) and superior temporal gyrus (i.e., Wernicke's area) have been commonly reported in ASD. Autistic patients show abnormal activation of parietal and occipital brain regions, associated with visual imagery, for comprehending sentences that do not invite visual imagery (Dichter 2012). This results suggest a processing strategy that more reliant on visualization to support language comprehension.

Studies of restricted and repetitive behaviors using fMRI are limited, realizing that head motion is a major source of error in data analysis. However, the cognitive manifestations of these features have been addressed using a variety of paradigms requiring cognitive control, including but not limited to spatial attention, working memory, interference inhibition, and response monitoring, revealing anomalous activation in frontostriatal brain regions(see (Dichter 2012) for review).

6.3.2 Functional Connectivity

fMRI data cannot in itself prove causality, and such data can only state if a certain brain region is involved in a specific task. The fMRI-based connectivity (fcMRI) studies assess interregional temporal correlations of the BOLD signal. Thus, fcMRI studies in ASDs addresses circuitry-level questions believed to be central to autistic symptoms. Growing evidence indicates predominantly reduced long-distance functional connectivity in ASD. For example, lower connectivity between frontal and posterior parietal and temporal cortical regions has been consistently reported. This network plays central roles in processing social-affective information.

Furthermore, a marked decrease in connectivity between the right cerebellar region and the supratentorial regulatory language areas, and between the Broca's regions and modulatory control dorsolateral prefrontal region, may underlie the abnormal language function in children with ASD (Just et al. 2012). Cortical–cortical under connectivity has been replicated with a variety of fMRI paradigms related to core symptom domains, such as language comprehension, social cognition, executive function, or motor tasks (for review see (Anagnostou and Taylor 2011; Hugdahl et al. 2012; Stigler et al. 2011; Williams and Minshew 2007; Just et al. 2012)).

6.3.3 Resting-State Functional Connectivity

Resting-state fMRI (rsfMRI) derives from the rationale that low frequency spontaneous oscillations in BOLD signal are temporally coherent among brain areas that are functionally and probably structurally connected. Thus, the resting-state network is believed to reflect a rudimentary and intrinsic organization of the resting brain, providing the opportunity to study long and short-range connectivity in ASD, without complications related to task activation. rsfMRI has been increasing applied to the study of the complex alterations of functional networks in ASD. Most of the results of such studies are consistent with decreased long-distance connectivity. There are reports of decreased frontal-posterior connectivity at rest in the default mode network in adolescents and adults with ASD (Weng et al. 2010). Correlations between social and repetitive behavior symptoms and a number of resting connectivity metrics in individuals with ASD have been reported (Dennis and Thompson 2013; Anagnostou and Taylor 2011; Dichter 2012).

More recent rsfMRI suggests that ASDs may also be characterized by aberrant increase in local functional connectivity. Posterior overconnectivity has been reported to be associated with higher ASD symptom severity (Keown et al. 2013). In the same way, local overconnectivity in occipital and posterior temporal regions appears consistent with the preference for local over global visual processing in ASD (Maximo et al. 2013). Monk et al. reported stronger functional connectivity between the posterior cingulate and the temporal lobe and parahippocampal gyrus in ASD (Monk et al. 2009). Whereas under connectivity reflects reduced efficiency of within-network communication in ASD, degrading the capacity of integrating information from widespread and diverse systems, increased local connectivity can be attributed to impaired experience-driven mechanisms, and could be involved in the development of the outstanding capacities in some within the autistic spectrum.

Ongoing efforts attempt to map resting-state connectivity using sleep fMRI. This approach would appear to be well suited to studying early emerging brain activity linked to speech processing in infant, where classical rsfMRI studies are unlikely. In summary, rsfMRI findings reflect important aspects of network dysfunction associated with socio-communicative, cognitive, and sensori motor impairments in ASD.

6.4 Magnetic Resonance Spectroscopy

Magnetic resonance spectroscopy (MRS) provides a noninvasive method to measure the chemical concentration of certain metabolites in brain tissues, characterize tissue metabolic processes in vivo and identify chemical and structural abnormalities associated with diseases. The most widely used and relevant nuclei for in vivo biomedical applications are hydrogen (^{1}H), phosphorus (^{31}P), and carbon (^{13}C). As nonproton MRS is still an emergent field and few studies have been reported in autism, the following section will focus on ^{1}H-MRS and the analysis of the most important metabolites studied in ASD: N-acetyl aspartate, glutamate/glutamine, creatine/phosphocreatine, choline-containing compounds, and myo-inositol. Representative spectra of ASD and TD, that show the peak values of the typical metabolites, are displayed in Fig. 6.4.

N-acetyl aspartate (NAA) is an amino acid that is present almost exclusively in neurons, neuronal projections, and mature oligodendrocytes. The NAA is considered as a marker of neuronal density and function and/or mitochondrial dysfunction, with the levels that increase during the early brain development until adulthood (van der Knaap et al. 1990). A large number of studies in children reliably reveal that ASD patient's exhibit reduced NAA levels in a wide number of cortical and subcortical brain regions. Consistently, reported areas with NAA reduction include global gray or matter; frontal cortex, anterior cingulate cortex, hippocampus-amygdala region, temporal and occipital lobes, thalamus, and cerebellum (for review see (Hyde et al. 2010; Raznahan et al. 2013; Anagnostou and Taylor 2011; Ipser et al. 2012;

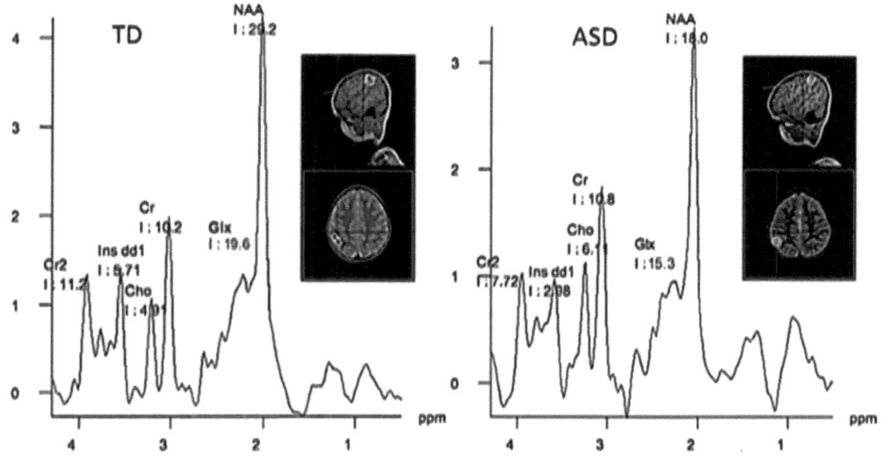

Fig. 6.4 Representative 1H-MRS-SVS spectra in the temporoparietal junction of a typical development (*TD*) subject and autism spectrum disorder (*ASD*) subject. Notice a reduced *NAA* and *Glx* and increased *Cho* concentrations in ASD. The inset depicts the three-dimensionally localized MRS voxel. Data courtesy of the Brain Images Processing Group at CIREN, Havana

Chugani 2012)). Decreased neuronal density would be the origin of disruptions in long-range fiber tracts, discussed in the next section, thus resulting in decreased NAA in both the grey and the corresponding white matter.

On the other hand, MRS at high fields allows for quantification of glutamate/glutamine compounds (Glx) addressing the role played by these excitatory metabolites in the neurochemistry balances in the ASD brain. MRS data from children cohorts indicate that autism groups exhibit reduced Glx levels in cortical gray matter (DeVito et al. 2007), anterior cingulate cortex (Bernardi et al. 2011), frontal and occipital cortex, and cerebellum (DeVito et al. 2007). Recent research have suggested elevated Glx levels in ASD in the anterior cingulate cortex (Bejjani et al. 2012) and putamen (Doyle-Thomas et al. 2014) of children, and the auditory cortex of adults (Brown et al. 2013). These results are consistent with the hyperglutamatergic theories of ASD, supporting an excitation/inhibition imbalance underlying the pathophysiology of ASD. Further researches are needed in larger cohorts of patients and solving methodological issues in the acquisition and processing of MRS spectra.

Metabolites' levels are often reported as ratios with regard to chromium (Cr) concentration, instead of absolute metabolite levels. The explicit rationale for this approach is that Cr is assumed to be stable in the normal, as well as in several pathologic states. Ratios correct for the partial volume effect and MRI inhomogeneities, thus preventing the intra and inter-subject variance and are critical in quantifying low-abundance metabolites. However, Cr level is roughly considered as a measure of cellular energy metabolism, serving as the storage and use mechanism for ATP synthesis. A decrease in absolute concentrations has been reported during hypermetabolic process, probably in relation with homeostasis of cellular bioenergetics (Mountford et al. 2010). Thus, the common practice of using creatine/phosphocreatine to calculate metabolite ratios may lead to confounding results in the study of ASD's biochemistry. Relative to controls, reductions in Cr levels have been reported in the occipital cortex (Ipser et al. 2012) and global gray matter for ASD children (Friedman et al. 2006). Ipser et al. reported in a meta-analysis based on 20 studies higher levels of Cr for ASD adults than children in global grey matter (Ipser et al. 2012). A trend toward increases in the temporal lobe in adults has been also suggested.

Further neurochemical imbalances were found in other ^{1}H-MRS studies, such as decreased concentrations of choline-containing compounds (Cho) and myoinositol (mI) suggesting decreased cellularity or density (Friedman et al. 2006). A recent study found correlations between Glx, Cho, and mI in caudate, putamen, and thalamus, and behavioral scores in ASD children (Doyle-Thomas et al. 2014). Additional studies will be needed to support these findings and elucidate the sensitivity of these metabolites to diagnosis of ASD.

Taken together, this evidence does not support the explanation that early structural overgrowth is due to an acceleration of the normal brain maturation processes. More likely, these findings are consistent with an excess of minicolumn density as described by Casanova and colleagues (Casanova and Trippe 2009; Casanova 2006). Alternative hypotheses include reduced synaptic density, poorly differentiated cortex, and possibly inflammatory processes (Friedman et al. 2006). Studies in adults provide further support that a typical change of neurochemical profile in ASD groups expands beyond childhood and persists during adulthood. Nonetheless, the conflation of age group, the overlap of diagnosis that meets the ASD criteria, the unmatched use of sedation, and the heterogeneity of regions studied, place limits on the strength of the conclusions that can be drawn from ^{1}H-MRS studies.

6.5 Diffusion Tensor Imaging

6.5.1 White Matter Organization

Diffusion tensor imaging (DTI) is sensitive to loss of white matter integrity and to differences in anatomical connectivity. Fractional anisotropy (FA) is a measurement that indirectly reflects the diameter and density of fibers, myelination and macrostructural features such as fiber coherence, while mean diffusivity (MD) is a measure of the mean motion of water considered in all directions. Thus, lower FA or higher MD suggests a disruption in microstructural brain white matter features.

Children and adolescents with autism show areas of abnormal white matter integrity relative to TD subjects. Reports of decreased FA included white matter structures in the frontal and left temporal lobes (Ke et al. 2009), inferior longitudinal fasciculus (ILF, comprising the right fusiform face area), superior longitudinal fasciculus (SLF), and corpus callosum/cingulum (Jou et al. 2011). Abnormalities in white matter organization in ASD have been found to persist into adulthood. Regions of decreased FA in the corpus callosum of the frontal and temporal lobes were reported in participants with age ranging from older children and adolescents to adults (Keller et al. 2007; Alexander et al. 2007). A wide meta-analysis including 25 DTI studies with regions of interest methods demonstrated consistently significant FA reductions in the corpus callosum, left uncinate fasciculus, and left superior longitudinal fasciculus, and significant increases of MD in the corpus callosum and superior longitudinal fasciculus bilaterally, in subjects with ASD compared with the TD individuals (Aoki et al. 2013). These findings have been taken as evidence that white matter abnormalities may constitute the biological basis of the decreased functional connectivity in adults with autism, discussed in Sect. 3.1. However, an apparent tendency to normalization of white matter maturation over time has been reported, possibly accounting for the behavioral improvements often observed in high-functioning autism (Bakhtiari et al. 2012).

6.5.2 Anatomical Connectivity

Diffusion tensor imaging (DTI) studies limited to voxel-wise comparisons of FA are useful for identifying the presence of white matter pathology in ASD, but are limited in its ability to identify the underlying fiber tracts involved. Several strategies have been implemented to achieve the tract reconstruction and connectivity analysis from DTI. The latest can be roughly divided into global or tract-based analysis, exemplified in Fig. 6.5. In the literature, a dominant finding with respect to the anatomical connectivity in ASD is that there is a combined pattern of local over-connectivity and long distance under-connectivity, illustrated for our data in Fig. 6.5b).

A growing body of the DTI literature examining ASDs provides evidence of impaired neural connectivity in the corpus callosum/cingulum and the temporal lobes affecting the major fiber tracts known to connect key nodes involved in emotional face processing, language, and executive functioning. The DTI studies have shown reduced structural connectivity between sensory cortices and prefrontal regions, visual areas and temporoparietal regions, as well as other regions identified as components of the multisensory network such as superior temporal sulcus, occipitotemporal junction, and anterior cingulate regions (Casanova 2006). More recent DTI-data support the notion that stronger short-range connectivity may develop in parallel with decreased long-range connectivity. Greater local connectivity in the ASD individuals have been reported in the left superior parietal lobule, the precuneus and angular gyrus, and the right supramarginal gyrus (Li et al. 2014).

Our data in individuals with ASD from ages 4–7 shows a disruption of long-range connections involving the superior longitudinal fasciculus (SLF) and uncinate fasciculus (Fig. 6.5c). SLF is associated with spatial working memory and the integration of the auditory and speech areas of the brain. The left SLF is important for information exchange between Broca's and Wernicke's areas. Thus, abnormal

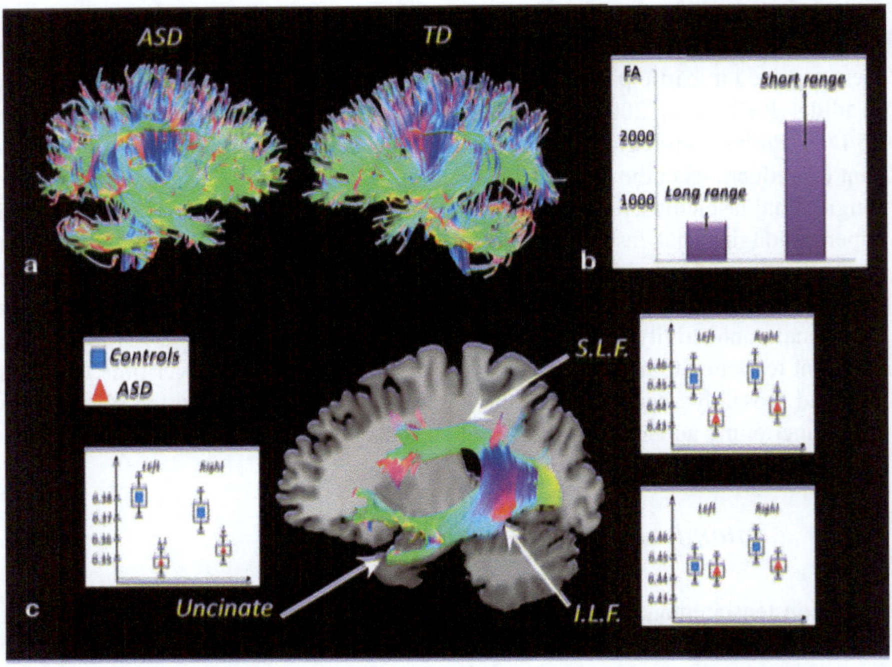

Fig. 6.5 *Upper panel*: example of global connectivity analysis. **a** Representative *DTI* reconstruction from an *ASD* subject and an age matched *TD* subject. The same reconstruction parameters, algorithms and thresholds were used. A reduction in fiber density and organization in ASD subject is noticeable. **b** Prevalence of short-range vs. long-range connectivity in ASD networks. *Lower panel*: example of tract-based analysis in 15 children with ASD and age matched controls. **c** Significant differences were found in the superior longitudinal fasciculus (*S.L.F*) and uncinate fasciculus, but not in the inferior longitudinal fasciculus (*I.L.F*). The figure has been compiled by the authors from several different studies conducted by the Brain Images Processing Group at CIREN

connectivity in SLF may underpin the neurobiological basis of language deficits in ASD. On the other hand, the uncinate fasciculus has traditionally been considered to be a part of the limbic system and is known for its involvement in human emotion processing, memory and language functions, all of which are impaired in ASD (Melillo and Leisman 2009).

6.6 Emission Tomography

Positron emission tomography (PET) and single-photon emission tomography (SPECT) provide sensitive means for tracking biochemical and molecular processes in ASD. Both are nuclear medicine imaging techniques that measure signal from radioisotopes intravenously injected into the blood stream. There are two basic approaches to examining the changes in brain function associated with ASD. First, resting state studies allow to detect regional abnormalities in brain metabolism, blood flow, and neuroreceptor binding. Second, activation studies can reveal abnormalities in the functional networks involved in motor, cognitive or visual task.

6.6.1 *Resting Metabolic Studies in ASD*

Measurement of regional glucose metabolism and regional cerebral blood flow (rCBF) secondarily reflect the underlying synaptic activity. Both children and adults with autism show decrease in rCBF in superior temporal areas, known to be involved in cognitive processing of complex sounds. The conclusion from these studies was that children and adults with ASD had a dysfunction in the temporal regions correlated with the severity of autistic disorders. Focal perfusion abnormalities in the left medial prefrontal cortex (BA 9 and BA 10) and in the anterior cingulate gyrus (BA 32) have been associated with the social components of autistic features. For an extensive review, see Ref. (Williams and Minshew 2007; Chugani 2012; Lauvin et al. 2012).

6.6.2 *Receptor Binding in ASD*

There is a huge increase in the number of labeled ligands available for neurotransmitter and neuroreceptor mapping using PET and SPECT. However, few have been used in the study of autism. These studies have focused in investigating the functional integrity of the dopaminergic, serotonergic, and gamma-aminobutyric acid (GABA-mediated) systems. While the efforts to elucidate the contribution of the neurotransmission systems in the pathophysiology of autism have failed to identify the neurochemical substrates of autism, they have provided convincing evidence of altered neurotransmission in both children and adults with autism.

The integrity of the dopaminergic system can be tested using PET with the tracer ^{18}F-fluorodopa (FDOPA). After its intravenous administration, FDOPA is taken up, metabolized and stored by the terminals of the dopaminergic projections. Ernst and colleagues reported a significant reduction in FDOPA uptake in the anterior medial prefrontal cortex in the autistic group compared to the control group (Ernst et al. 1997). The dopamine transporter binding can be studied using ^{99m}Tc-TRODAT-1 imaged by SPECT or ^{11}C-WIN-35,428 imaged with PET. Increase in the dopamine transporter binding has been reported throughout the whole brain in children, and in orbital frontal cortex in adults with autism (Chugani 2012).

Abnormal findings in serotonergic system have been consistently reported in ASD. Pioneering work from Chugani and colleagues provided evidence of focal abnormalities in serotonin synthesis along the dentate-thalamo-cortical pathway in autistic children, using α-^{11}C-methyl-L-tryptophan imaged with PET (Chugani et al. 1997). Decreased serotonin transporter binding has been reported throughout medial frontal cortex, midbrain, and temporal lobes of children with autism, determined by the SPECT radiotracer ^{123}I-nor-B-CIT (Makkonen et al. 2008). Similarly, a PET study using ^{11}C-McN-5652 detected reduction in serotonin transporter in the anterior and posterior cingulate cortices of autistic adults, correlated with impairment in social cognition, whereas the reduction of binding in the thalamus was correlated with repetitive or obsessive behavior (Nakamura et al. 2010). Murphy and colleagues studied the serotonergic neurotransmission using the SPECT tracer ^{123}I-5-I-R91150 for receptor binding in a group with Asperger's syndrome (Murphy et al. 2006). They found significantly reduced serotonin receptor binding in total, anterior, and posterior cingulate cortex, bilateral in frontal and superior temporal lobes and in the left parietal lobe. In summary, nuclear medicine imaging techniques provide convincing evidence of altered serotonin synthesis, transporter and receptors in small cohorts of children and adults with autism.

The symptoms of ASD have been linked to alterations in GABA neurotransmission system. Recent preliminary investigation used emission tomography studies to address this hypothesis. Mendez and colleagues found significantly lower ^{11}C-Ro15-4513 binding imaged with PET in adults with ASD (Mendez et al. 2013), in accordance with a decrease in the accumulation of ^{123}I-IMZ in the superior and medial frontal cortex measured by SPECT (Mori et al. 2012). These results provide initial evidence of a GABA(A) deficit in ASD and support further investigations of the GABA system in this disorder.

6.6.3 Activation Studies

Similar to fMRI, neurovascular coupling forms the basis for mapping patterns of activation in the working human brain using PET and SPECT. Considering that increase in regional blood flow are subtle, obtaining a reliable activation map depends on sophisticated subtraction algorithms, which calculate the differences between the pattern of blood flow or metabolism during the task and the control state. Emission tomographies exhibit less temporal and spatial resolution than fMRI. Nevertheless, its capacity to display the brain activity at the time of the uptake in the tissue makes it a suitable alternative to obtain valuable information about task-related changes in brain functioning of noncooperative patients, such as autistic children.

An early study of autistic male patients using ^{15}O-water PET reported a reduced level of activation of the right dentate nucleus and the left frontal area in response to language stimuli (Rapoport and Cutler 1985). The same radiotracer was used to explore the brain organization of audition and language in high-functioning autistic adults (Muller et al. 1999). The results provided support for a typical hemispheric dominance for receptive language in autism, probably related with delay in maturation of structures that normally participates in expressive language, cognitive-attentional and affective functions. Other PET studies have reported abnormal pattern of cortical activation in individuals with autism. Metabolism of temporal regions, specialized in the perception and integration of complex sounds, is affected in both children and adults with ASD. These results were interpreted as evidence that the left hemispheric language processing areas do not function efficiently in autism. The impairment in the activation of the so-called "theory of mindh network (superior temporal sulcus, medial frontal gyrus, right temporal pole) in ASD was demonstrated using PET with stimuli of animations of triangles that performed interacted movements that evoked social interpretations. An extensive review can be seen in (Williams and Minshew 2007).

Few case studies have been carried out in individuals with savant skills. These studies found increased activation across cortical regions, especially over frontal and parietal lobes, in the savant as compared with typical controls (Corrigan et al. 2012). Although the number of studies is limited, they provide useful evidence that in some individuals with autism, certain overdeveloped brain circuits may lead to extraordinary abilities.

6.7 Autism as a Distributed Disorder

Autism is a disorder primarily affecting high-order integration processes, in the social, emotional, linguistic, motor, attentional, executive, and visuospatial domains. As described throughout this chapter, neuroimaging studies support substantial evidence to suggest the notion of neural network reorganization in ASD. In this framework, it is obvious that ASD cannot be explained as a localized defect, but needs to be modeled as a network disorder (Just et al. 2012; Melillo and Leisman 2009). This has led researchers to focus on the characterization of large-scale brain connectivity in order to move the knowledge of factors underlying ASD further ahead.

In sections 3.1 and 4.2 were outlined the fMRI and DTI-based techniques to address functional and anatomical connectivity, respectively. Results of these approaches in ASD point to altered brain connectivity as a key feature of its pathophysiology. In fact, the cortical underconnectivity theory emerged from fMRI studies that showed impairment in the coordinated functioning between frontal and posterior brain regions in autism. This under-synchronization has been replicated in a wide variety of task using fMRI and electroencephalography (EEG). DTI studies seem to suggest that impairments in white matter integrity may underlie the reduced coordination of activity across brain regions, giving rise to core features of ASD.

The development of a unifying model of connectivity in ASD has been challenged by the heterogeneity and sometime contradictory results. Most of the data available are consistent with decreased long-distance connectivity, but increased short-range connectivity has also been reported. Rather than simply considering these findings as contrary, mixed results from neuroimaging suggest different neural networks architecture in the brains of individuals with ASD, where pathological connections probably coexist with compensatory reorganization of anatomical and functional networks. Recently, multicenter studies have been designed to identify ASD progression biomarkers to improve the understanding of disease neurobiology. For example, the Autism Brain Imaging Data Exchange (ABIDE) comprised MRI, rsfMRI, and phenotypic information from 539 individuals with ASD and 573 age-matched TD controls. Preliminary results from ABIDE highlight the core significance of the insula, posterior cingulate cortex, and thalamus. Moreover, whole-brain

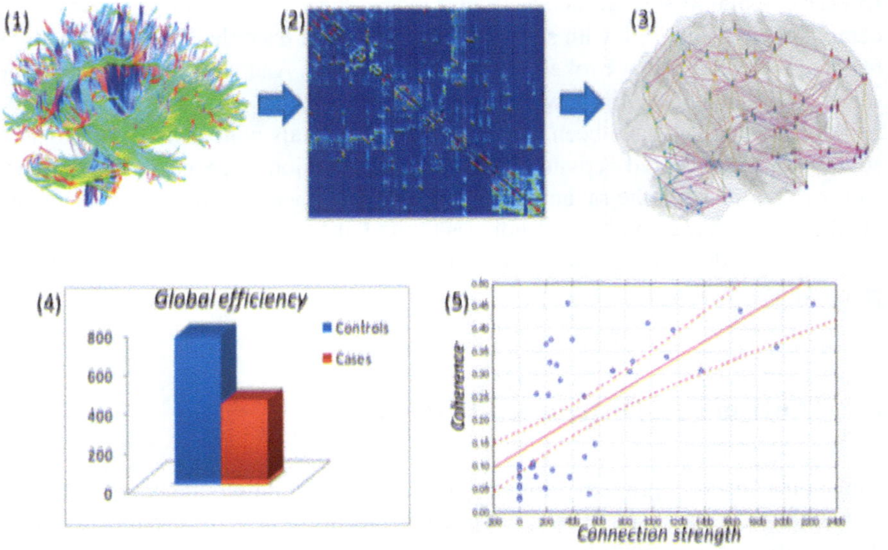

Fig. 6.6 *Upper*: General scheme for the extraction of the network metrics. *1* DTI-based connectivity networks; *2* Connectivity matrices are extracted using connectivity strengths between anatomical regions; *3* Whole brain network visualized in anatomical space. A node in the network, depicted as a *dot*, represents a brain region. The connection strength between two regions is depicted by a *line*. *Lower: 4* Comparison of global efficiency for ASD (*red*) and TD (*blue*) groups. *5* Significant correlation between anatomical and functional connectivity in a graph theory framework. Data courtesy of the Brain Images Processing Group at CIREN, Havana

analyses reconciled seemingly disparate findings of both hypo- and hyperconnectivity in ASD, with expected dominance of hypoconnectivity, particularly for cortico–cortical and interhemispheric functional connectivity (DiMartino et al. 2013).

Recent approaches have applied graph theory to anatomic and functional connectivity data derived from neuroimaging, in order to study whole-brain network topology. Typical graph theoretical approaches treat individual voxels or regions as nodes in a complex network and assess the structural or functional properties of how these nodes interact in a dynamical system (Fig. 6.6, 1–3). After the whole brain connectivity networks are constructed, the network topology is investigated and statistical analysis is employed to compare ASD with TD using network metrics and connectivity strengths between regions. Disturbances of connectivity in ASD involve significant changes of network parameters such as the global efficiency and characteristic path length, as measures of network integration, and cluster coefficient and local efficiency, as measures of local specialization. Decreased global efficiency in ASD networks, as shown in Fig. 6.6 (4), indicates a disruption in the capacity of the network to maintain efficient global information flow between its nodes. Higher cluster coefficients, smaller characteristic path length in functional networks (Hai et al. 2014), and differences in global and local wiring cost in morphological networks (Ecker et al. 2013)have also been reported in ASD in comparison with TD groups.

However, the relation between abnormal circuitry and the behavioral patterns observed in ASD is not straightforward. A recent study showed that standard EEG coherence analysis did not resemble structural connectivity in autism (Coben et al. 2014). More advanced approaches to functional and anatomical analysis may provide more detailed and accurate information about the complicated picture of connectivity anomalies in ASD. Based on innovative statistical approaches to EEG coherence analysis (Machado et al. 2014) and graph theory, we were able to predict the behavior of lower EEG bands from DTI data in ASD, shown in Fig. 6.6 (5). This suggest that integration of higher bands activities (e.g., gamma band), associated with a variety of integrative processes including attention, face processing, and emotional arousal, with anatomical circuitry into a unified approach, may require more sophisticated models. In this context, graph theory can provide a common framework to deal with the complexity of dynamic networks in ASD, as it addresses simultaneously principles of brain functioning such as specialization and integration. Thus, atypical patterns of network topology could become hallmark neurobiological features of autism in the future.

6.8 Conclusions

Although a diagnostic imaging marker is at present still lacking, recent neuroimaging techniques reveal significant functional and microstructural changes in the brains of autistic children and adults. This highlights the importance of a synergistic and integrated use of multiple imaging modalities to gain new insight into the pathophysiology of ASD, diagnosis of which remains clinical. Imaging techniques have the potential of elucidating the anatomical and functional substrate of aberrant connectivity in autism and as such hold promise to ultimately assist with the development of diagnostic markers and biologically guided treatments in ASD. Available data suggest that the neural networks in the brains of individuals with ASD are connected and organized differently than the neural networks in the brain of TD individuals. Inasmuch, the investigation of interregional connectivity is necessary for the efficient completion of higher cognitive functions; the introduction of causality analysis and complex networks approaches could represent the basis for the modification of current image-based diagnostic procedures in ASD. In summary, neuroimaging provides a tantalizing glimpse into the neurophysiologic basis of ASD, revealing valuable information for devising new strategies for further investigation of these disorders.

References

Aleman-Gómez Y, Melie L, Valdes P (2006) IBASPM: toolbox for automatic parcellation of brain structures. Presented at the 12th Annual Meeting of the Organization for Human Brain Mapping, Florence, Italy. Available on CD-Rom in NeuroImage, vol 27, no 1

Alexander AL, Lee JE, Lazar M, Boudos R, DuBray MB, Oakes TR et al (2007) Diffusion tensor imaging of the corpus callosum in autism. Neuroimage 34:61–73

Anagnostou E, Taylor MJ (2011) Review of neuroimaging in autism spectrum disorders: what have we learned and where we go from here. Mol Autism 2:4

Aoki Y, Abe O, Nippashi Y, Yamasue H (2013) Comparison of white matter integrity between autism spectrum disorder subjects and typically developing individuals: a meta-analysis of diffusion tensor imaging tractography studies. Mol Autism 4:25

Bakhtiari R, Zurcher NR, Rogier O, Russo B, Hippolyte L, Granziera C et al (2012) Differences in white matter reflect atypical developmental trajectory in autism: a tract-based Spatial Statistics study. Neuroimage Clin 1:48–56

Brambilla P, Hardan A, di Nemi S, Perez J, Soares J, Barale F (2003) Brain anatomy and development in autism: review of structural MRI studies. Brain Res Bull 61:557–569

Bejjani A, O'Neill J, Kim JA, Frew AJ, Yee VW, Ly R et al (2012) Elevated glutamatergic compounds in pregenual anterior cingulate in pediatric autism spectrum disorder demonstrated by 1H MRS and 1H MRSI. PLoS One 7:e38786

Bernardi S, Anagnostou E, Shen J, Kolevzon A, Buxbaum JD, Hollander E et al (2011) In vivo 1H-magnetic resonance spectroscopy study of the attentional networks in autism. Brain Res 1380:198–205

Brieber S, Neufang S, Bruning N, Kamp-Becker I, Remschmidt H, Herpertz-Dahlmann B et al (2007) Structural brain abnormalities in adolescents with autism spectrum disorder and patients with attention deficit/hyperactivity disorder. J Child Psychol Psychiatr 48:1251–1258

Brown MS, Singel D, Hepburn S, Rojas DC (2013) Increased glutamate concentration in the auditory cortex of persons with autism and first-degree relatives: a (1)H-MRS study. Autism Res 6:1–10

Brun CC, Nicolson R, Lepore N, Chou YY, Vidal CN, DeVito TJ et al (2009) Mapping brain abnormalities in boys with autism. Hum Brain Mapp 30:3887–3900

Casanova MF (2006) Neuropathological and genetic findings in autism: the significance of a putative minicolumnopathy. Neuroscientist 12:435–441

Casanova M, Trippe J (2009) Radial cytoarchitecture and patterns of cortical connectivity in autism. Philos Trans R Soc Lond B Biol Sci 364:1433–1436

Cauda F, Costa T, Palermo S, D'Agata F, Diano M, Bianco F et al (2014) Concordance of white matter and gray matter abnormalities in autism spectrum disorders: a voxel-based meta-analysis study. Hum Brain Mapp 35:2073–2098

Chen R, Jiao Y, Herskovits EH (2011) Structural MRI in autism spectrum disorder. Pediatr Res 69:63R–68R

Chugani DC (2012) Neuroimaging and neurochemistry of autism. Pediatr Clin North Am 59:63–73

Chugani DC, Muzik O, Rothermel RD, Behen M, Chakraborty PK, Mangner TJ et al (1997) Altered serotonin synthesis in the dentatothalamocortical pathway in autistic boys. Ann Neurol 42:666–669

Chung MK, Dalton KM, Alexander AL, Davidson RJ (2004) Less white matter concentration in autism: 2D voxel-based morphometry. Neuroimage 23:242–251

Coben R, Mohammad-Rezazadeh I, Cannon RL (2014) Using quantitative and analytic EEG methods in the understanding of connectivity in autism spectrum disorders: a theory of mixed over- and under-connectivity. Front Hum Neurosci 8:45

Corrigan NM, Richards TL, Treffert DA, Dager SR (2012) Toward a better understanding of the savant brain. Compr Psychiatr 53:706–717

Courchesne E, Karns CM, Davis HR, Ziccardi R, Carper RA, Tigue ZD et al (2001) Unusual brain growth patterns in early life in patients with autistic disorder: an MRI study. Neurology 57:245–254

Dennis EL, Thompson PM (2013) Typical and atypical brain development: a review of neuroimaging studies. Dialogues Clin Neurosci 15:359–384

DeVito TJ, Drost DJ, Neufeld RW, Rajakumar N, Pavlosky W, Williamson P et al (2007) Evidence for cortical dysfunction in autism: a proton magnetic resonance spectroscopic imaging study. Biol Psychiatr 61:465–473

Dichter GS (2012) Functional magnetic resonance imaging of autism spectrum disorders. Dialogues Clin Neurosci 14:319–330

DiMartino A, Ross K, Uddin L, Sklar AB, Castellanos FX, Milham MP (2009) Functional brain correlates of social and non-social processes in autism spectrum disorders: an ALE meta-analysis. Biol Psychiatr 65:63–74

DiMartino A, Yan CG, Li Q, Denio E, Castellanos FX, Alaerts K et al (2014) The autism brain imaging data exchange: towards a large-scale evaluation of the intrinsic brain architecture in autism. Mol Psychiatr 19:659–667

Doyle-Thomas KA, Card D, Soorya LV, Wang AT, Fan J, Anagnostou E (2014) Metabolic mapping of deep brain structures and associations with symptomatology in autism spectrum disorders. Res Autism Spectr Disord 8:44–51

Ernst M, Zamatkin AJ, Matochik JA (1997) Low medial prefrontal dopaminergic activity in autistic children. Lancet 350:638

Ecker C, Ronan L, Feng Y, Daly E, Murphy C, Ginestet CE et al (2013) Intrinsic gray-matter connectivity of the brain in adults with autism spectrum disorder. Proc Natl Acad Sci U S A 110:13222–13227

Friedman SD, Shaw DW, Artru AA, Dawson G, Petropoulos H, Dager SR (2006) Gray and white matter brain chemistry in young children with autism. Arch Gen Psychiatr 63:786–794

Hai L, Xue Z, Ellmore TM, Frye RE, Wong ST (2014) Network-based analysis reveals stronger local diffusion-based connectivity and different correlations with oral language skills in brains of children with high functioning autism spectrum disorders. Hum Brain Mapp 35:396–413

Hardan AY, Jou RJ, Keshavan MS, Varma R, Minshew NJ (2004) Increased frontal cortical folding in autism: a preliminary MRI study. Psychiatr Res 131:263–268

Hardan AY, Muddasani S, Vemulapalli M, Keshavan MS, Minshew NJ (2006) An MRI study of increased cortical thickness in autism. Am J Psychiatr 163:1290–1292

Hardan AY, Libove RA, Keshavan MS, Melhem NM, Minshew NJ (2009) A preliminary longitudinal magnetic resonance imaging study of brain volume and cortical thickness in autism. Biol Psychiatr 66:320–326

Hazlett HC, Poe M, Gerig G, Styner M, Chappell Ch, Smith RG et al (2011) Early brain overgrowth in autism associated with an increase in cortical surface area before age 2. Arch Gen Psychiatr 68:467–476

Heeger DJ, Rees D (2002) What does fMRI tell us about neuronal activity? Nat Neurosci 3:142–151

Hugdahl K, Beyer MK, Brix M, Ersland L (2012) Autism spectrum disorder, functional MRI and MR spectroscopy: possibilities and challenges. Microb Ecol Health Dis 23:18960

Hyde KL, Samson F, Evans AC, Mottron L (2010) Neuroanatomical differences in brain areas implicated in perceptual and other core features of autism revealed by cortical thickness analysis and voxel-based morphometry. Hum Brain Mapp 31:556–566

Ipser JC, Syal S, Bentley J, Adnams CM, Steyn B, Stein DJ (2012) 1H-MRS in autism spectrum disorders: a systematic meta-analysis. Metab Brain Dis 27:275–287

Jiao Y, Chen R, Ke X, Chu K, Lu Z, Herskovits EH (2010) Predictive models of autism spectrum disorder based on brain regional cortical thickness. Neuroimage 50:589–599

Jou RJ, Jackowski AP, Papademetris X, Rajeevan N, Staib L, Volkmar FR (2011) Diffusion tensor imaging in autism spectrum disorders: Preliminary evidence of abnormal neural connectivity. Aust N Z J Psychiatr 45:153–162

Just MA, Keller TA, Malave V, Kana RK, Varma R (2012) Autism as a neural systems disorder: a theory of frontalposterior underconnectivity. Neurosci Biobehav Rev 36:1292–1313

Ke X, Tang T, Hong S, Hang Y, Zou B, Li H et al (2009) White matter impairments in autism, evidence from voxel-based morphometry and diffusion tensor imaging. Brain Res 1265:171–177

Keller TA, Kana RK, Just MA (2007) A developmental study of the structural integrity of white matter in autism. Neuroreport 18:23–27

Keown CL, Shih P, Nair A, Peterson N, Mulvey ME (2013) Local functional overconnectivity in posterior brain regions is associated with symptom severity in autism spectrum disorders. Cell Rep 5:567–572

Kwon H, Ow AW, Pedatella KE, Lotspeich LJ, Reiss AL (2004) Voxel-based morphometry elucidates structural neuroanatomy of high-functioning autism and Asperger syndrome. Dev Med Child Neurol 46:760–764

Lauvin MA, Martineau J, Destrieux C, Andersson F, Bonnet-Brilhault F, Gomot M et al (2012) Functional morphological imaging of autism spectrum disorders: current position and theories proposed. Diagn Interv Imag 93:139–147

Li H, Xue Z, Ellmore TM, Frye RE, Wong ST (2014) Network-based analysis reveals stronger local diffusion-based connectivity and different correlations with oral language skills in brains of children with high functioning autism spectrum disorders. Hum Brain Mapp 35:396–413

Machado C, Estevez M, Leisman G, Melillo R, Rodriguez R, Defina P et al (2014) QEEG spectral and coherence assessment of autistic children in three different experimental conditions. J Autism Dev Disord. doi:10.1007/s10803-013-1909-5

Makkonen I, Riikonen R, Kokki H, Airaksinen MM, Kuikka JT (2008) Serotonin and dopamine transporter binding in children with autism determined by SPECT. Dev Med Child Neurol 50:593–597

Maximo JO, Keown CL, Nair A, Muller RA (2013) Approaches to local connectivity in autism using resting state functional connectivity MRI. Front Hum Neurosci 7:605

Melillo R, Leisman G (2009) Autistic spectrum disorders as functional disconnection syndrome. Rev Neurosci 20:111–131

Mendez MA, Horder J, Myers J, Coghlan S, Stokes P, Erritzoe P et al (2013) The brain GABA-benzodiazepine receptor alpha-5 subtype in autism spectrum disorder: a pilot [(11)C]Ro15-4513 positron emission tomography study. Neuropharmacology 68:195–201

Mori T, Mori K, Fujii E, Toda Y, Miyazaki M, Harada M et al (2012) Evaluation of the GABAergic nervous system in autistic brain: (123)I-iomazenil SPECT study. Brain Dev 34:648–654

Monk CS, Peltier SJ, Wiggins JL, Weng SJ, Carrasco M, Risi S et al (2009) Abnormalities of intrinsic functional connectivity in autism spectrum disorders. Neuroimage 47:764–772

Mountford C, Stanwell P, Lin A, Ramadan S, Ross B (2010) Neurospectroscopy: the past, present and future. Chem Rev 110:3060–3086

Muller RA, Behen M, Rothermel RD, Chugani DC, Muzik O, Mangner TJ et al (1999) Brain mapping of language and auditory perception in high-functioning autistic adults: a PET study. J Autism Dev Disord 29:19–31

Murphy DG, Daly E, Schmitz N, Toal F, Murphy KC, Curran S et al (2006) Cortical serotonin 5-HT2A receptor binding and social communication in adults with Asperger's syndrome: an in vivo SPECT study. Am J Psychiatr 163:934–936

Nakamura K, Sekine Y, Ouchi Y, Tsujii M, Yoshikawa E, Futatsubashi M et al (2010) Brain serotonin and dopamine transporter bindings in adults with high-functioning autism. Arch Gen Psychiatr 67:59–68

Nickl-Jockschat T, Habel U, Michel TM, Manning J, Laird AR, Fox PT et al (2012) Brain structure anomalies in autism spectrum disorder–a meta-analysis of VBM studies using anatomic likelihood estimation. Hum Brain Mapp 33:1470–1489

Nordahl CW, Dierker D, Mostafavi I, Schumann CM, Rivera SM, Amaral DG et al (2007) Cortical folding abnormalities in autism revealed by surface-based morphometry. J Neurosci 27:11725–11735

Rapoport SI, Cutler NR (1985) Brain metabolism in autism. Arch Gen Psychiatr 42:448–455

Raznahan A, Lenroot R, Thurm A, Gozzi M, Hanley A, Spence SJ et al (2013) Mapping cortical anatomy in preschool aged children with autism using surface-based morphometry. Neuroimage Clin 2:111–119

Rojas DC, Peterson E, Winterrowd E, Reite ML, Rogers SJ, Tregellas JR (2006) Regional gray matter volumetric changes in autism associated with social and repetitive behavior symptoms. BMC Psychiatr 6:56

Schumann CM, Bloss CS, Barnes CC, Wideman GM, Carper RA, Akshoomoff N et al (2010) Longitudinal magnetic resonance imaging study of cortical development through early childhood in autism. J Neurosci 30:4419–4427

Stigler KA, McDonald BC, Anand A, Saykin AJ, McDougle CJ (2011) Structural and functional magnetic resonance imaging of autism spectrum disorders. Brain Res 1380:146–161

van der Knaap MS, van der Grond J, van Rijen PC, Faber JA, Valk J, Willemse K (1990) Age-dependent changes in localized proton and phosphorus MR spectroscopy of the brain. Radiology 176:509–515

Weng SJ, Wiggins JL, Peltier SJ, Carrasco M, Risi S, Lord C et al (2010) Alterations of resting state functional connectivity in the default network in adolescents with autism spectrum disorders. Brain Res 1313:202

Williams DL, Minshew NJ (2007) Understanding autism and related disorders: what has imaging taught us? Neuroimaging Clin N Am 17:495–509

Chapter 7
Autism, Development and Neural Plasticity

Maria de los Angeles Robinson-Agramonte, Maria Elena Gonzàlez Fraguela and Jorge Bergado-Rosado

Abstract In this chapter, we show evidences derived from carefully designed animal experiments and clinical data, supporting a role of neuroplasticty on the development and physiopathology of autism. Although the exact mode in which genetic and environmental factors interact in various psychiatric conditions, including autism remains largely unclear, a comprehensive understanding of these complex interactions, including the contribution of immune molecules, in the establishment of neuronal connectivity that may drive into phenotypes of autism spectrum disorder (ASD) is of the greatest importance to understand its causes and develop successful strategies to treat and revert its consequences. Available evidence strongly supports the involvement of maladaptive neuroplasticity, mutations, epigenetic modification, and individual miRNA-related pathways in the pathogenesis and pathophysiology of the complex clinical phenotypes that is included in ASD.

Keywords Autism · Neural plasticity · Neurogenetics · Cellular mechanism · Neural development · Neuroimmunlogy · Connectivity

---

M. d. l. A. Robinson-Agramonte
Neuroimmunology Department, International Center for Neurological Restoration, Havana, Cuba
e-mail: maria.robinson@infomed.sld.cu

M. E. Gonzàlez Fraguela
Neurobiochemistry Lab, Department of Immunochemistry, International Center for Neurological Restoration, Havana, Cuba
e-mail: marie@neuro.ciren.cu

J. Bergado-Rosado (✉)
Department of Experimental Neurophisiology, International Center for Neurological Restorations, Ave 25 # 15805, 11300 Havana, Cuba
e-mail: bergado@neuro.ciren.cu

M. d. l. A. Robinson-Agramonte (ed.), *Translational Approaches to Autism Spectrum Disorder*, DOI 10.1007/978-3-319-16321-5_7

7.1 Introduction

Psychiatric illnesses, including autism spectrum disorder (ASD) are disabling disorders with a poorly understood pathophysiology. However, it is becoming increasingly evident that these disorders result from the disruptions across whole cellular networks, including the cellular plasticity cascades, rather than any particular monoamine system, which lead to aberrant information processing and to altered functions and diseases. Alterations of synaptic connectivity have been well documented as cause of the aberrant social behavioral and other main symptoms in autism. Our objective in this chapter is to offer clues towards a better understanding of all these events in autism and how it could reveal fundamental insights into the causes of these devastating illnesses and/or offer potential novel targets for the intervention.

In its modern form, autism was first described independently by Kanner (Kanner 1943) in the USA and Asperger (Asperger 1944) in Germany in the fourth decade of the twentieth century. Both men were born in the Austro–Hungarian Empire, and they described what are now considered two of the main forms of autism within the ASD: low- and high-functioning autism. Some consider that Asperger might have been autistic himself (Lyons and Fitzgerald 2007). Initially considered a psychological disturbance, clinical and experimental evidences have led to the current concept that ASD is the result of a generalized neurodevelopment disorder.

The molecular and cellular mechanisms involved in the development of the nervous system are the same that sustain the fundamental property of neural plasticity (Bergado-Rosado and Almaguer-Melian 2000). Neural plasticity can be defined as the ability of the nervous system to change its functional and structural properties in response to external or internal challenges (Bergado-Rosado and Almaguer-Melian 2000). Structural changes include neurite extension (axons or dendrites), the eventual formation or elimination of synapses, as well as (at least in certain regions of the adult brain) to breed new neurons. Functional plasticity implicates mainly changes in the efficacy of synaptic transmission in existing synapses, both in the form of an increase (i.e., long-term potentiation (LTP)) or a decrease (i.e., long-term depression (LTD)).

7.2 Neural Plasticity

The concept of neural plasticity evolved during the twentieth century from a very rigid and immutable vision of the nervous system (Cajal 1911) to one in constant change and remodeling due to what in general terms we call experience (Greenough et al. 1987). Important landmark in this conceptual evolution was the demonstration of the live experiences that (i.e., laboratory rats living in enriched environments) shape the brain after birth (Rosenzweig et al. 1962a, b). Similarly, the demonstrated influences of visual inputs (i.e., monocular deprivations in kitten) in the postnatal development of the visual system (Hubel and Wiesel 1998) contribute to break the wall of immutability. The proven relationship between neuronal metabolism and memory consolidation (Flood et al. 1975; Flexner et al. 1965) and the recognition of memory as the most common expression of plasticity in the adult nervous system (Matthies 1976, 1989), reminded us that neurons, beyond cable properties, are living cells and not simple devices in electric circuits. The experiments performed by Merzenich and coworkers in the 1980s (Merzenich et al. 1984, 1983) provided irrefutable evidence that even in the adult brain, well-established circuits—like those expressed in the sensory and motor homunculi—were able to change under the drive of experience. Finally, the demonstration in birds that new neurons could be generated in certain areas of the adult brain (Nottebohm 1989, 2011) rapidly extended to mammals, signed the birth certificate of a new way to understand the nervous system.

At the cellular level, neuroplastic changes rely on the mechanisms that are present since earlier periods of neural development like neurogenesis, to other that express in later stages, i.e., neurite grow and synaptogenesis, and finally changes in synaptic connectivity that occur during the whole life (for example LTP and LTD).

The cellular machinery involved in the regulation and expression of these processes includes the participation of an external messenger, represented by a neurotransmitter, a hormone, a growth factor or a component of the extracellular matrix. Any of these agents interact with specific receptors linked to molecular cascades in which protein kinases play the essential role. These enzymes can act at two different levels: in the cytoplasm, where they phosphorylate existing proteins modifying their functional status and in the nucleus, where kinases act as transcription factors that lead to the activation of genes and the synthesis of new proteins. These proteins determine the functional and structural modifications that can result in neuritic extension or retraction, synapse formation or elimination, or modifications in synaptic efficacy.

LTP was demonstrated in 1973 (Bliss and Lomo 1973) and is a good example of how this control machinery works to produce experience-driven changes in neural function. The classic form of LTP is induced by high-frequency stimulation of excitatory glutamatergic afferents to a neuronal population, allowing the entrance of a calcium current into the postsynaptic neurons through N-methyl-D-aspartate (NMDA) receptors (Collingridge 1985; Collingridge and Bliss 1987; Dupuis et al. 2014; Collingridge et al. 2013, 1992; Bashir et al. 1991; Volianskis et al. 2013). Increased intracellular calcium activates kinases like the calcium–calmodulin-activated kinase (CaMK) (Hedou and Mansuy 2003; Huber et al. 1995a, b, c) which is anchored to the postsynaptic densities (Zhang and Lisman 2012). Activated CaMK phosphorilates α-amino-3-hydroxy-5-methyl-4-isoxazolepropionic acid (AMPA) receptors (Andrasfalvy and Magee 2004; Derkach et al. 1999; Lisman and Zhabotinsky 2001) and promotes the membrane inclusion of vesicles (called endosomes) which carry AMPA receptors to the postsynaptic membrane (Lisman and Zhabotinsky 2001; Hayashi et al. 2000) that are presumably responsible for an increased membrane response to glutamate.

This mechanism is able to sustain a temporary increase of synaptic reactivity in what is now conceived as an early phase of LTP (E-LTP) (Matthies et al. 1990). For a longer maintenance of LTP (more than 4 h), the synthesis of new proteins is required (Frey et al. 1988; Krug et al. 1984). This phase of late LTP (L-LTP) is carried out by the combined action of protein synthesis and distribution to the potentiated synapses by a mechanism of synaptic tagging demonstrated by Frey and Morris in the late 1990s (Frey and Morris 1997a, b).

LTP-related protein synthesis can be activated by different kinases. Some, like CaMK, are activated by glutamatergic mechanisms; whereas others like protein kinase A (PKA) require the cooperative activation of cathecholaminergic synapses. They are part of the cascades whose final products enter the nucleus and promote the synthesis of proteins required to maintain and consolidate the change in synaptic efficacy, collectively termed as *plasticity-related proteins* (McNair et al. 2006; Nedivi et al. 1993). The subunits of AMPA receptors are among them.

After the formation of the neural tube and the neural crests, a chain of processes transform this single-layered structure into the spinal cord and our convoluted brain (Pescosolido et al. 2012). The first step in this transition relies on successive mitotic divisions that generate trillions of neural cells (including glia). Much more neurons are generated than those that remain in the brain. A process of selective death via apoptosis eliminates as many as the 50 % of the new formed nerve cells before birth. Genetic mutations that interfere with neural apoptosis are lethal (Jessell and Sanes 2013).

Many of the new generated neurons migrate following the routes marked by radial glial to form layered cortices. Other follow the route marked by attraction or repulsion factors until they reach their final destination and enter an intense phase of differentiation (Valiente and Marín 2010). Mutations affecting the proteins controlling migration can lead to severe malformation like is the case of lissencephaly (Marín and Rubenstein 2003; Valiente and Marín 2010), a condition associated with severe mental retardation and seizures. Both processes are independent from experience, and are mainly controlled by genetic factors activated by local substrates (i.e., positional value in the egg).

Neurite elongation is an essential part of neural differentiation. Axons can extend only a few microns in the case on local interneuron's, or dozens of centimeters in the case of projection neurons. Axon growth, all the way to their target, results from the interaction of receptors expressed on the tip of the axon (the growth cone) and markers that can attract or repel this "exploring" device (Huber et al. 2003; Rossi et al. 2007).

Synaptogenesis occurs later in development, when axons reach a target and find a dendritic counterpart in the form of a filopodium (Scheiffele 2003). The initial contact forms a bond which is later reinforced mechanically by trans-synaptic binding molecules, and functionally by the formation of vesicles, a docking complex, and a postsynaptic density, among other (Scheiffele 2003; McAllister 2007). This initial synapse contacts are guided mainly by factors that do not depend on experience. The process of neurite growth and synaptogenesis are not perfect. Most of the axons do reach their main target, but others travel to distant regions. Consequently, many synapses initially formed are eliminated (synaptoapotosis) (Mattson et al. 1998) during the last stages of intrauterine development and the early period of postnatal life. However, this selective pruning is not perfect and many of these "aberrant" synaptic connections remain. Those are the base of the so-called *cross-modal plasticity* (i.e., the visual cortex process auditory signals in blinded persons) (Bavelier and Hirshorn 2010)or the intriguing phenomenon of synaesthesia (i.e., a sensation of color is evoked by tasting a flavor) (Sinner 2012).

The process of refinement of synaptic connections depends on experience that is characterized by Greenough as experience-expectant plasticity (Greenough et al. 1987). During the early years of postnatal development, the nervous system undergoes maturational changes. Some depend from the kind, intensity, and timing of stimuli that are common to the members of one species, but need to be present at the right time to occur. The main mechanism is the already mentioned process of synaptic pruning of many aberrant synapses and the functional reinforcement of other connections leading to the acquisition of improved sensory and motor abilities (perception, walking, manipulative skills) and language. The classical experiments of Hubel and Weisel in monocularly deprived kittens are excellent examples of this process (Hubel and Wiesel 1964). The need of external stimuli for this maturation to occur, along with the need that these stimuli occur at the right time, leads to the concept of *critical periods of development* (Berardi et al. 2000).

Autism is now believed to be a consequence of failures of these early postnatal maturational processes (LeBlanc and Fagiolini 2011; Berger et al. 2013), and as it happens in many other instances affecting development, it can have genetic causes, but the contribution of an inadequate experience should not be overseen.

7.3 ASD and Developmental Mechanisms

Genetic factors have been demonstrated in conditions like the Fragile X syndrome, tuberous sclerosis, and Rett disease that share some behavioral traits with autism and are now considered within ASD. In fact, autism is a highly heritable condition, as shown by familiar and twin studies (Banerjee et al. 2014; Goldani et al. 2014; Anderson 2014). However, the diversity of conditions under the ASD is so wide that it seems very difficult to identify a small number of affected genes. In recent reviews, more than 20 genes linked to ASD are mentioned (Banerjee et al. 2014; Pescosolido et al. 2012). Interestingly, many of them are directly linked to neural development, synaptogenesis, and synaptic function.

An interesting approach could be the study of gene networks using new computational tools like the so-called weighted gene co-expression network analysis (WGCNA). Ben David and Shifman applied WGCNA to genetic data provided by the Allen Human Brain Atlas project (http://www.brain-map.org) and were able to identify modules of co-expressed genes that contain the genes potentially related to some forms of ASD. One of the modules contained genes that produce proteins essential for neural plasticity, while the other is enriched in proteins required for synapse formation and function (Ben-David and Shifman 2012a, 2013, 2012b).

Epigenetic modifications of gene expression have also been implicated in ASD. These mechanisms suppress expression of genes by the methylation of DNA or acetylation of the proteins (histones) that cover the DNA molecule (Mehler 2008). Epigenetic modifications are important in neural development, (Fagiolini et al. 2009; Mathers and McKay 2009; O'Shea 2001) neural function, and plasticity (Borrelli et al. 2008; Covic et al. 2010; Day and Sweatt 2011; Levenson and Sweatt 2005; Lubin 2011; Lubin et al. 2011; Puckett and Lubin 2011) as well as in neuropathology (Denk and McMahon 2012; Feinberg 2007; Hwang et al. 2013; Lee and Ryu 2010; Qureshi and Mehler 2010; Smith 2011; Tsankova et al. 2007). It has been suggested that a dysregulation of DNA methyltransferases and histone deacetylases may reduce the repression of genes, leading to an aberrant pattern of connectivity between relevant brain areas in autistic subjects (Mbadiwe and Millis 2013).

The recently discovered microRNAs (miRNAs) are a class of endogenous small noncoding RNAs able to regulate translation or transcription of specific mRNAs (Liu and Kohane 2009; Xu et al. 2012). They can repress translation of hundreds of genes and target a multitude of cellular mechanisms (Costa-Mattioli et al. 2009; Costa et al. 2012). miRNAs expressed in the brain are implicated in maintaining normal neuronal function and homeostasis which in turn is associated with memory, neuronal differentiation, synaptic plasticity, and neurogenesis as well as neuronal degeneration (Barry 2014; Costa-Mattioli et al. 2009; Krichevsky et al. 2003; Schratt et al. 2006). Thus, dysfunctional miRNAs might underlay the disturbed synaptic connectivity in ASD.

A dysregulation in the expression of two miRNAs of the same family (132/232), important for the experience-dependent cortical plasticity, has been the most relevant finding in ASD. The *miR-132* also regulates the immune response and is recognized to be induced by endotoxines (Im and Kenny 2012; Hansen et al. 2013) and to be affected by cytomegalovirus infection, contributing to autism pathogenesis (Onore et al. 2012b, a). Also, it is known that *miR-132* has an effect on the synaptic structure, which is modulated by the Fragile X protein (FMRP), a product of the *Fmr* gene responsible for fragile X mental retardation, and a reduced level of *miR-132* is found in Rett syndrome (D'Hulst and Kooy 2009; Xu et al. 2012; Olde Loohuis et al. 2012).

Other miRNAs, like *miR-195*, *miR-219*, *miR-181b*, *miR-128b*, and *miR-346* may underlay an altered cascade of molecular events leading to maladaptive neural plasticity and autism (Chan and Kocerha 2012; Edmonson et al. 2014; Ziats and Rennert 2013b, a; Sarachana et al. 2010). *miR-195* is known to target brain-derived neurotrophic factor (BDNF) (Mellios and Sur 2012) and *miR-128b* to inhibit cAMP response element-binding (CREB) (Lit et al. 2012), whereas *miR132/232* are induced by BDNF and positively regulate CREBS signaling, a transcription factor that regulates diverse cellular responses, including proliferation, survival, and differentiation (Hollander et al. 2010; Iacoangeli et al. 2010). Other presumed miRNAs linked to autism are *miR-495* and *miR-381*, and two BDNF-regulating miRNAs (Cristino et al. 2014; Sarachana et al. 2010; Wu and Levitt 2013; Edmonson et al. 2014; Lit et al. 2012).

While the majority of miRNAs upregulated by both LTP and LTD, temporal expression patterns seem to differ between these two forms of plasticity (Park and Tang 2009). So, results from Goodman and cols in a knockout (KO) mouse lacking *miR-132* and/or *Mr-212* showed the involvement of both miRNAs in the maturation of dendrites of newborn neurons in the adult hippocampus, while simultaneous deletion of the *miR-132/212* genes caused a severe reduction in dendritic length, arborization, and spine density. On a similar road miR-132 inhibitors were able to block CREB effects on dendritic maturation, arguing the suggestion about the involvement of *miR-132* in CREB signaling (Magill et al. 2010). It is also important to point that BDNF-induced *miR-132* upregulates glutamate receptors at the synapse, as a support to the notion on *miR-132* role in controlling synaptic structure and function (Gao et al. 2010; Olde Loohuis et al. 2012).

Different pathways can modulate miR-132's effects on synaptic plasticity and connectivity to contribute autism pathology. Briefly, mutations in methyl CpG binding protein 2 (Mecp2) increase *miR-132* expression via BDNF and by mir-137 binding its promoter transcriptional inhibition (Peca and Feng 2012; Wright et al. 2013). However, the major relevance in autism may rest on *miR-132/212* induced by BDNF and regulating positively CREB signaling (Hollander et al. 2010) in combination with *miR-495* and *miR-381* also regulated by BDNF. Other relevant pathway influenced by an altered miRNAs expression in ASD is the phosphoinositide 3-kinase (P13K)/protein kinase B (PKB/Akt), also affected by BDNF signaling. This PKB is induced by phosphatase and tensin homolog (PTEN), a controller of Akt, and PTEN mutations are responsible for a high percentage of subtypes of syndromic ASD. Akt is activated by trophic factors and once that trophic factor binds its specific receptor, P13K is recruited and an anti-apoptotic effect may occur by direct regulation of the apoptotic pathway and cellular metabolism and by transcriptional control of molecules promoting cell survival (Huang and Tindall 2007; Song et al. 2005; Vaishnavi et al. 2013).

Results obtained in animal models of Fragile X syndrome suggest that this genetic condition and ASD also share a deficient gamma-aminobutyric acidergic (GABAergic) inhibition, and consequently, excess excitability in critical areas like the amygdala, the cortex (Paluszkiewicz et al. 2011), and the hippocampus. There is also growing evidence that "a dysfunction of the GABAergic signaling early in development leads to a severe E/I unbalance in neuronal circuits, a condition that may account for some of the behavioral deficits observed in ASD patients" (Pizzarelli and Cherubini 2011). Early postnatal development is a period in which experience shape and adjust the nervous system to its mature, functional basic form, and a balance between excitation and inhibition is essential for the correct outcome of critical maturation processes (LeBlanc and Fagiolini 2011).

Several important "players" in neural plasticity molecular cascades have been suggested to have causal links with conditions within the ASD. In animal models of autism, an elevated level of NMDA receptors and enhanced LTP (Rinaldi et al. 2007)or impaired LTD (Moretti et al. 2006) have been described.

As mentioned before, protein kinases are essential links between external signals and nuclear regulation. One pleiotropic serine–threonine kinase called *m*ammalian *t*arget *of r*apamycin (mTOR) has gained attention for its central role in plasticity and its presumed relationships to autism. mTOR is influenced by activation of glutamate receptors like the NMDAR, the AMPAR, and the metabotropic glutamate receptor

(GluR), acting as a connecting node to downstream kinases like phoshoinositide dependent kinase, phosphatidyl inositol kinase, Akt, and tuberous sclerosis complex protein 1 and 2 (Hoeffer and Klann 2009). Altered mTOR-signaling impairs synaptic plasticity and possibly other form of neural plasticity because of an altered protein synthesis and turnover. Dysregulated mTOR activity has been documented in tuberous sclerosis, neurofibromatosis, Fragile X, and other neurological disease like autism, all inserted in a complex network of molecular pathway involving epigenetics (i.e., immune factors) and genetic factors linked to neural plasticity (Gipson and Johnston 2012).

7.4 Autism Development and Immune Mediators

Neuroinflammation, a CNS-specific inflammation that do not reproduce the classic characteristics of inflammation in the periphery, can engender progressive events similar to neurodegenerative disorders (Morgan et al. 2010; Pardo et al. 2005; Vargas et al. 2005). In general, neuroinflammation is characterized by the reactivity of microglial cells with an increased expression and/or release of cytokines and chemokines and other neurotoxic factors (Monnet-Tschudi et al. 2011; Zimmerman et al. 2005; Vargas et al. 2005; Chez et al. 2007), and unbalance of neurotrophins (NGF, BDNF, NT3/4) reported as altered in autism. Several studies have provided strong support to the theory of neuroinflammation occurring at disperse brain regions like: the cerebellum, midfrontalcortex, Brodmann's Area 40 (BA 40) in the parietal cortex, BA9 (superior frontal cortex), and cingulate gyrus (Pardo et al. 2005; Vargas et al. 2005; Zimmerman et al. 2005) in ASD. Neurotrophic factors and their receptors like BDNF and TrkB are necessary for plasticity and able to induce plasticity (Bramham and Messaoudi 2005; Cao et al. 2013).

The nervous system share numerous immune molecules; among these are human leukocyte antigen (HLA) genes located within a large genomic region named as major histocompatibility complex (MHC). These immune molecules play integral roles in the CNS throughout neural development, affecting neurogenesis, neuronal migration, axon guidance, synapse formation, activity-dependent refinement of circuits, and synaptic plasticity (Garay and McAllister 2010; Deverman and Patterson 2009). The prevailing hypothesis is that immune molecules in neurons engage MHC receptors, generating common intracellular signals in both neurons and immune cells that may result in an altered synaptic strength, neuronal morphology, and circuit properties downstream of synaptic activity (Ohtsuka et al. 2008; Syken et al. 2006; Fourgeaud et al. 2010; Fourgeaud and Boulanger 2010).

Earlier studies suggested that mothers of children with ASD share more HLA haplotype than typically developing mother–child pairs. Further studies showed that a particular HLA (HLA-DR4) occurs more often in children with ASD. Hypothesis-driven genetic assessments revealed that the HLA A2 (MHC class I) and HLA DR4 (MHC class II) loci contained single nucleotide polymorphisms (SNPs) associated with ASD, reinforcing the evidences on the association of genomic variation in the MHC region with ASD (Stubbs et al. 1985; Stubbs 1981; Torres et al. 2002, 2006; Hu et al. 2006).

Other data have suggested that genetic abnormalities in the MHC in autism included the genes located in the proximity of HLA genes like MET (encoding tyrosine kinase), the serine ad threonine kinase C gene PRKCB, PTEN mentioned above and the reelin gene (*RELN*) as well as elevated levels of inflammatory cytokines in the CNS contributing to behavioral pattern in ASD have been also reported (Hu et al. 2006).

It has been hypothesized that MHC class I acts at CNS synapses to mediate activity-dependent synaptic weakening and elimination of inappropriate connections, whereas alternative hypothesis suggest an indirect contribution to neural refinement by altering the structure and function of synapses and changing glutamate-receptor-mediated synaptic activity patterns (Datwani et al. 2009; Shatz 2009). Nevertheless, further studies are required to establish the roles for MHC molecules in refinement of connections during CNS development and their impact in neurodevelopmental disorders.

7.5 Autism and Connectivity

The cellular mechanisms underlying activity-dependent refinement of connections during CNS development are thought to involve a regulated combination of Hebbian synaptic plasticity, LTP/LTD and synaptic pruning, and homeostatic plasticity (Fourgeaud et al. 2010; Fourgeaud and Boulanger 2010).

In the later-developing cortical regions, severe disruptions resulted in a diminished functional connectivity among the regions of autistic brains (Verly et al. 2014a, b; Wass 2011; Ziats and Rennert 2013b). Long-distance underconnectivity is reported in many studies showing an abnormal brain connectivity associated to an upregulated GABA neurotransmission (Di et al. 2011; Kana et al. 2009; Wass 2011; Minshew and Keller 2010; Just et al. 2007), whereas others show a lower functional connectivity between the higher-order working memory/executive areas and the visuospatial regions at frontal and parietal–occipital level demonstrated by a combination of behavioral, functional magnetic resonance imaging (fMRI), functional connectivity, and corpus callosum morphometry (Damarla et al. 2010, Pate et al. 2010). A diminished connectivity between anterior and posterior insula and other specific brain regions involved in emotional and sensory processing has also been reported in ASD (Ebisch et al. 2011). Impaired connectivity between nearly all striatal subregions and heteromodal associative and limbic cortex also has been implicated in the physiopathology of ASD (e.g., insular and right superior temporal gyrus) (Di et al. 2011) as well as at level of fiber tract abnormalities in the corpus callosum, internal capsule, and middle cerebellar peduncle and all three segments of the internal capsule (Courchesne 1997; Ziats and Rennert 2013b). The cerebellum strongly implicated in the connectivity process, shows anatomic abnormalities in autism and the Purkinje cell, the main output cell in the cerebellum, is also significantly diminished in number in this disorder (Courchesne 1991, 1997; Belmonte et al. 2004).

On the other hand, studies performed by Kova et al. (Kovacs et al. 2009) show that neurological and non-neurological symptoms, like impaired social interaction, impaired verbal communication, failure to respond name, repetitive movements, avoidance of eye contact with other people, unusually sensitive to light, sound and touch, self-abusive behavior, restless movement, poor judgment, changes in mood and personality, oblivious to pain as well as increase gut permeability, all described in autism are a consequence of an altered synaptic transmission (Mor et al. 2011; Dijkstra et al. 2004; de Theije et al. 2011, 2014a, b).

The consistent pattern emerging across several studies is that while intrinsic functional connectivity in adolescents and adults with autisms is generally reduced, compared with age-matched controls, functional connectivity in younger children with the disorder appears to be increased (Uddin et al. 2013). Reports using electroencephalography recordings show that the greatest alterations of the dynamic brain connectivity are present in children with the worse autistic symptoms (Valko et al. 2007; Streit et al. 2004; Rodriguez and Kern 2011). More details are given in the chapter of electrophysiology.

7.6 Conclusions

While the importance of genetics—from genes to miRNAs—molecular cascades, supports a heritable of ASD, the particular role of early postnatal neural plasticity and its shaping by experience are not well understood. Studies are required to assess the impact of parental care, social interactions, affection, and other sources of experience during infancy. These early ages are associated with maturative transformations of the nervous system that might be heavily impaired if not properly stimulated. That is what is called a critical developmental period, and evidence of their impact in children development is old: mental retardation as a consequence of hospitalism or anaclitic depression (Crandall 1897), amblyopia resulting in children with congenital cataracts (Webber and Wood 2005), or language inabilities in the few document cases of feral children (Abello 1970). In the case of ASD, there are still more questions than answers; can a wrong nurture, by itself provoke autism? Or the contribution of a genetic background is mandatory? Is it becoming a popular belief that an early and intensive exposure of children to TV can cause autism, and indirect statistical data offer some support to such view (Waldman et al. 2006).

Unraveling the links between nature and nurture in ASD is essential for the prevention and treatment of this devastating condition, since targets from the molecular cascade involving genetic factor might be promissory and open new powerful therapeutic strategies.

References

Abello, VB. (1970) Wolf children: truth or fallacy? Clin Pediat 9:425–429

Anderson, GM. (2014) Autism biomarkers: challenges, pitfalls and possibilities. J Autism Dev Disord

Andrasfalvy BK, Magee JC (2004) Changes in AMPA receptor currents following LTP induction on rat CA1 pyramidal neurons. J Physiol 559(Pt 2):543–554

Asperger H (1944) Die "Autistichen Psychopaten" im Kindesalter. Arch Psychiat Nervkrank 117:76–136

Banerjee S, Riordan M, Baht MA (2014) Genetic aspects of autism spectrum disorders: insights from animal models. Front Cell Neurosci 8:58

Barry G (2014) Integrating the roles of long and small non-coding RNA in brain function and disease. Mol Psychiatry 19:410–416

Bashir ZI, Alford S, Davies SN, Randall AD, Collingridge GL (1991) Long-term potentiation of NMDA receptor-mediated synaptic transmission in the hippocampus. Nature 349:156–158

Bavelier D, Hirshorn EA (2010) I see where you're hearing: how cross-modal plasticity may exploit homologous brain structures. Nat Neurosci 13:1309–1311

Belmonte MK, Allen G, Beckel-Mitchener A, Boulanger LM, Carper RA, Webb SJ (2004) Autism and abnormal development of brain connectivity. J Neurosci 24:9228–9231

Ben-David E, Shifman S (2012a) Networks of neuronal genes affected by common and rare variants in autism spectrum disorders. PLoS Genet 8:e1002556

Ben-David E, Shifman S (2012b) Networks of neuronal genes affected by common and rare variants in autism spectrum disorders. PLos Genet 8(3):e1002556

Ben-David E, Shifman S (2013) Combined analysis of exome sequencing points toward a major role for transcription regulation during brain development in autism. Mol Psychiatry 18:1054–1056

Berardi N, Pizzorusso T Maffei L (2000) Critical periods during sensory development. Curr Opin Neurobiol 10:138–145

Bergado-Rosado JA, Almaguer-Melian W (2000) Cellular mechanisms of neuroplasticity. Rev Neurol 31:1074–1095

Berger JM, Rohn TT, Oxford JT (2013) Autism as the early closure of a neuroplastic critical period normally seen in adolescence. Biol Syst Open Access 1. doi:10.4172/2329-6577.1000118

Bliss TV, Lomo T (1973) Long-lasting potentiation of synaptic transmission in the dentate area of the anaesthetized rabbit following stimulation of the perforant path. J Physiol (Lond) 232:331–356

Borrelli E, Nestler EJ, Allis CD, Sassone-Corsi P (2008) Decoding the epigenetic language of neuronal plasticity. Neuron 60:961–974

Bramham CR, Messaoudi E (2005) BDNF function in adult synaptic plasticity: the synaptic consolidation hypothesis. Progr Neurobiol 76:99–125

Cajal SR (1911) Hystologie du systeme nerveux de l'homme et des vertebres. Maloine, Paris.

Cao C, Rioult-Pedotti MS, Migani P, Yu CJ, Tiwari R, Parang K, Spaller MR, Geobel DJ, Marshall J (2013) Impairment of TrkB-PSD-95 signaling in Angelman syndrome. PLoS Biol 11(2):e1001478

Chan AW, Kocerha J (2012) The path to microRNA therapeutics in psychiatric and neurodegenerative disorders. Front Genet 3:82. doi:10.3389/fgene.2012.00082. (eCollection@2012)

Chez MG, Dowling T, Patel PB, Khanna P, Kominsky M (2007) Elevation of tumor necrosis factor-alpha in cerebrospinal fluid of autistic children. Pediatr Neurol 36:361–365

Collingridge GL (1985) Long-term potentiation in the hippocampus: mechanism of initiation and modulation by neurotransmitters. Trends Pharmacol Sci 6:407–411

Collingridge GL, Bliss TVP (1987) NMDA receptors—their role in long-term potentiation. Trends Neurosci 10:288–293

Collingridge GL, Randall AD, Davies CH, Alford S (1992) The synaptic activation of NMDA receptors and Ca2+ signalling in neurons. Ciba Found Symp 164:162–171; (discussion 172-5)

Collingridge GL, Volianskis A, Bannister N, France G, Hanna L, Mercier M, Tidball P, Fang G, Irvine MW, Costa BM, Monaghan DT, Bortolotto ZA, Molnar E, Lodge D, Jane DE (2013) The NMDA receptor as a target for cognitive enhancement. Neuropharmacology 64:13–26 doi:10.1016/j.neuropharm.2012.06.051. (Epub@2012 Jul 11)
Costa MC, Leitao AL, Enguita FJ (2012) Biogenesis and mechanism of action of small non-coding RNAs: insights from the point of view of structural biology. Int J Mol Sci 13:10268–10295
Costa-Mattioli M, Sossin WS, Klann E, Sonenberg N (2009) Translational control of long-lasting synaptic plasticity and memory. Neuron 61:10–26
Courchesne E (1991) Neuroanatomic imaging in autism. Pediatrics 87:781–790
Courchesne E (1997) Brainstem, cerebellar and limbic neuroanatomical abnormalities in autism. Curr Opin Neurobiol 7:269–278
Covic M, Karaca E, Lie DC (2010) Epigenetic regulation of neurogenesis in the adult hippocampus. Heredity 105:122–134
Crandall FM (1897) Hospitalism. Arch Pediat 14:448–454
Cristino AS, Williams SM, Hawi Z, An JY, Bellgrove MA, Schwartz CE, Costa LF, Claudianos C (2014) Neurodevelopmental and neuropsychiatric disorders represent an interconnected molecular system. Mol Psychiatry 19:294–301
D'hulst C, Kooy RF (2009) Fragile X syndrome: from molecular genetics to therapy. J Med Genet 46:577–584
Damarla SR, Keller TA, Kana RK, Cherkassky VL, Williams DL, Minshew NJ, Just MA (2010) Cortical underconnectivity coupled with preserved visuospatial cognition in autism: evidence from an fMRI study of an embedded figures task. Autism Res 3:273–279
Datwani A, Mcconnell MJ, Kanold PO, Micheva KD, Busse B., Shamloo M, Smith SJ, Shatz CJ (2009) Classical MHCI molecules regulate retinogeniculate refinement and limit ocular dominance plasticity. Neuron 64:463–470
Day JJ, Sweatt JD (2011) Cognitive neuroepigenetics: a role for epigenetic mechanisms in learning and memory. Neurobiol Learn Mem 96:2–12
De Theije CG, Wu J, Da Silva SL, Kamphuis PJ, Garssen J, Korte SM, Kraneveld AD (2011) Pathways underlying the gut-to-brain connection in autism spectrum disorders as future targets for disease management. Eur J Pharmacol 668(Suppl 1):S70–S80 doi:10.1016/j.ejphar.2011.07.013. (Epub@2011 Jul 27)
De Theije CG, Bavelaar BM, Lopes Da Silva S, Korte SM, Olivier B, Garssen J, Kraneveld AD (2014a) Food allergy and food-based therapies in neurodevelopmental disorders. Pediatr Allergy Immunol 25, 218–226
De Theije CG, Wopereis H, Ramadan M, Van ET, Lambert J, Knol J, Garssen J, Kraneveld AD, Oozeer R (2014b) Altered gut microbiota and activity in a murine model of autism spectrum disorders. Brain Behav Immun 37:197–206 doi:10.1016/j.bbi.2013.12.005. (Epub@2013 Dec 11)
Denk F, Mcmahon SB (2012) Chronic pain: emerging evidence for the involvement of epigenetics. Neuron 73:435–444
Derkach V, Barria A, Soderling TR (1999) Ca2+/calmodulin-kinase II enhances channel conductance of alpha-amino-3-hydroxy-5-methyl-4-isoxazolepropionate type glutamate receptors. Proc Natl Acad Sci USA 96:3269–3274
Deverman BE, Patterson PH (2009) Cytokines and CNS development. Neuron 64:61–78
Di MA, Kelly C, Grzadzinski R, Zuo XN, Mennes M, Mairena MA, Lord C, Castellanos FX, Milham MP (2011) Aberrant striatal functional connectivity in children with autism. Biol Psychiatry 69:847–856
Dijkstra G, Van GH, Jansen PL, Moshage H (2004) Targeting nitric oxide in the gastrointestinal tract. Curr Opin Investig Drugs 5:529–536
Dupuis JP, Ladepeche L, Seth H, Bard L, Varela J, Mikasova L, Bouchet D, Rogemond V, Honnorat J, Hanse E, Groc L (2014) Surface dynamics of GluN2B-NMDA receptors controls plasticity of maturing glutamate synapses. EMBO J 33:842–861
Ebisch SJ, Gallese V, Willems RM, Mantini D, Groen WB, Romani GL, Buitelaar JK, Bekkering H (2011) Altered intrinsic functional connectivity of anterior and posterior insula regions in high-functioning participants with autism spectrum disorder. Hum Brain Mapp 32:1013–1028

Edmonson C, Ziats MN Rennert OM (2014) Altered glial marker expression in autistic post-mortem prefrontal cortex and cerebellum. Mol Autism 5:3–5

Fagiolini M, Jensen CL, Champagne FA (2009) Epigenetic influences on brain development and plasticity. Curr Opin Neurobiol 19:207–212

Feinberg AP (2007) Phenotypic plasticity and the epigenetics of human disease. Nature 447:433–440

Flexner LB, Flexner JB, De La Haba G, Roberts RB (1965) Loss of memory as related to inhibition of cerebral protein synthesis. J Neurochem 12:535–541

Flood JF, Bennett EL, Orme E, Rosenzweig MR (1975) Relation of memory formation to controlled amounts of brain protein synthesis. Physiol Behav 15:97–102

Fourgeaud L, Boulanger LM (2010) Role of immune molecules in the establishment and plasticity of glutamatergic synapses. Eur J Neurosci 32:207–217

Fourgeaud L, Davenport CM., Tyler CM, Cheng TT, Spencer M.B. & Boulanger LM (2010) MHC class I modulates NMDA receptor function and AMPA receptor trafficking. Proc Natl Acad Sci USA 107:22278–22283

Frey U, Krug M, Reymann KG, Matthies H (1988) Anisomycin, an inhibitor of protein synthesis, blocks late phases of LTP phenomena in the hippocampal CA1 region in vitro. Brain Res 452:57–65

Frey U, Morris RGM (1997a) Synaptic tagging and long-term potentiation. Nature 385:533–536

Frey U, Morris RGM (1997b) Synaptic tagging: implications for late maintenance of hippocampal long-term potentiation. Trends Neurosci 21:181–188

Gao J, Wang WY, Mao YW, Graff J, Guan JS, Pan L, Mak G, Kim D, Su SC, Tsai LH (2010) A novel pathway regulates memory and plasticity via SIRT1 and miR-134. Nature 466:1105–1109

Garay PA, Mcallister AK (2010) Novel roles for immune molecules in neural development: implications for neurodevelopmental disorders. Front Synaptic Neurosci 2:136 doi:10.3389/fnsyn.2010.00136. (eCollection@2010)

Gipson TJ, Johnston MV (2012) Plasticity and mTOR: towards restoration of impaired synaptic plasticity in mTOR-Related neurogenetic disorders. Neural Plast 2012:486402

Goldani AA, Downs SR, Widjaja F, Lawton B, Hendren RL (2014) Biomarkers in autism. Front Psychiatry 5:100 doi:10.3389/fpsyt.2014.00100. (eCollection@2014)

Greenough WT, Black JE, Wallace CS (1987) Experience and brain development. Child Dev 58:539–559

Hansen KF, Karelina K, Sakamoto K, Wayman GA, Impey S, Obrietan K (2013) miRNA-132: a dynamic regulator of cognitive capacity. Brain Struct Funct 218:817–831

Hayashi Y, Shi SH, Esteban JA, Piccini A, Poncer JC, Malinow R (2000) Driving AMPA receptors into synapses by LTP and CaMKII: requirement for GluR1 and PDZ domain interaction. Science 287:2262–2267

Hedou G, Mansuy IM (2003) Inducible molecular switches for the study of long-term potentiation. Philos Trans R Soc Lond B Biol Sci 358 797–804

Hoeffer CA, Klann E (2009) mTOR signaling: at the crossroads of plasticity, memory and disease. Trends Neurosci 33:67–75

Hollander JA, Im HI, Amelio AL, Kocerha J, Bali P, Lu Q, Willoughby D, Wahlestedt C, Conkright MD, Kenny PJ (2010) Striatal microRNA controls cocaine intake through CREB signalling. Nature 466:197–202

Hu VW, Frank BC, Heine S, Lee NH, Quackenbush J (2006) Gene expression profiling of lymphoblastoid cell lines from monozygotic twins discordant in severity of autism reveals differential regulation of neurologically relevant genes. BMC Genomics 7:118

Huang H, Tindall DJ (2007) Dynamic FoxO transcription factors. J Cell Sci 120:2479–2487

Hubel DH, Wiesel TN (1964) Effects of monocular deprivation in kittens. Naunyn Schmiedebergs Arch Exp Pathol Pharmakol 248:492–497

Hubel DH, Wiesel TN (1998) Early exploration of the visual cortex. Neuron 20:401–412

Huber AB, Kolodkin AL, Ginty DD, Cloutier J-F (2003) Signaling at the growth cone: ligand-receptor complexes and the control of axon growth. Annu Rev Neurosci 26:509–563

Huber KM, Mauk MD, Kelly PT (1995a) Distinct LTP induction mechanisms: contribution of NMDA receptors and voltage-dependent calcium channels. J Neurophysiol 73:270–279

Huber KM, Mauk MD, Kelly PT (1995b) LTP induced by activation of voltage-dependent Ca2+ channels requires protein kinase activity. Neuroreport 6:1281–1284

Huber KM, Mauk MD, Thompson C, Kelly PT (1995c) A critical period of protein kinase activity after tetanic stimulation is required for the induction of long-term potentiation. Learn Mem 2:81–100

Hwang JY, Aromolaran KA, Zukin RS (2013) Epigenetic mechanisms in stroke and epilepsy. Neuropsychopharmacology 38:167–182

Iacoangeli A, Bianchi R, Tiedge H (2010) Regulatory RNAs in brain function and disorders. Brain Res 1338:36–47 doi:10.1016/j.brainres.2010.03.042. (Epub@2010 Mar@20)

Im HI, Kenny PJ (2012) MicroRNAs in neuronal function and dysfunction. Trends Neurosci 35:325–334

Jessell TM, Sanes JR (2013) Differentiation and survival of nerve cells. In: Kandel ER, Schwartz JH, Jessell TM, Siegelbaum SA, Hudspeth AJ (eds) Principles of neural science, 5th edn. McGraw Hill Medical, New York, pp 2618–2664

Just MA, Cherkassky VL, Keller TA, Kana RK, Minshew NJ (2007) Functional and anatomical cortical underconnectivity in autism: evidence from an FMRI study of an executive function task and corpus callosum morphometry. Cereb Cortex 17:951–961

Kana RK, Keller TA, Cherkassky VL, Minshew NJ, Just MA (2009) Atypical frontal-posterior synchronization of Theory of Mind regions in autism during mental state attribution. Soc Neurosci 4:135–152

Kanner L (1943) Autistic disturbances of affective contact. Nerv Child 2:217–250

Kovacs R, Rabanus A, Otahal J, Patzak A, Kardos J, Albus K, Heinemann U, Kann O (2009) Endogenous nitric oxide is a key promoting factor for initiation of seizure-like events in hippocampal and entorhinal cortex slices. J Neurosci 29:8565–8577

Krichevsky AM, King KS, Donahue CP, Khrapko K, Kosik KS (2003) A microRNA array reveals extensive regulation of microRNAs during brain development. RNA 9:1274–1281

Krug M, Lössner B, Ott T (1984) Anisomycin blocks the late phase of long-term potentiation in the dentate gyrus of freely moving rats. Brain Res Bull 13:39–42

Leblanc JJ, Fagiolini M (2011) Autism: A "Critical Period" Disorder? Neural Plast 2011

Lee J, Ryu H (2010) Epigenetic modification is linked to Alzheimer's disease: is it a maker or a marker? BMB Rep 43:649–655

Levenson JM, Sweatt JD (2005) Epigenetic mechanisms in memory formation. Nat Rev Neurosci 6:108–118

Lisman JE, Zhabotinsky AM (2001) A model of synaptic memory: a CaMKII/PP1 switch that potentiates transmission by organizing an AMPA receptor anchoring assembly. Neuron 31:191–201

Lit L, Sharp FR, Bertoglio K, Stamova B, Ander BP, Sossong AD. Hendren RL (2012) Gene expression in blood is associated with risperidone response in children with autism spectrum disorders. Pharmacogenomics J 12:368–371

Liu H, Kohane IS (2009) Tissue and process specific microRNA-mRNA co-expression in mammalian development and malignancy. PLoS ONE 4:e5436

Lubin FD (2011) Epigenetic gene regulation in the adult mammalian brain: multiple roles in memory formation. Neurobiol Learn Mem 96:68–78

Lubin FD, Gupta S, Parrish R.R., Grissom NM, Davis RL (2011) Epigenetic mechanisms: critical contributors to long-term memory formation. Neuroscientist 17:616–632

Lyons V, Fitzgerald M (2007) Did Hans Asperger (1906–1980) have Asperger Syndrome. J Autism Dev Disord 37:2020–2021

Magill ST, Cambronne XA, Luikart BW, Lioy DT, Leighton BH, Westbrook GL, Mandel G, Goodman RH (2010) microRNA-132 regulates dendritic growth and arborization of newborn neurons in the adult hippocampus. Proc Natl Acad Sci USA 107:20382–20387

Marín O, Rubenstein JLR (2003) Cell migration in the forebrain. Annu Rev Neurosci 26:483

Mathers JC, Mckay JA (2009) Epigenetics–potential contribution to fetal programming. Adv Exp Med Biol 646:119–123
Matthies H (1976) Cellular mechanisms of learning processes and the shaping of memory. Z Psychol 184:308–328
Matthies H (1989) In search of cellular mechanisms of memory. Progr Neurobiol 32:277–349
Matthies H, Frey U, Reymann K, Krug M, Jork R, Schroeder H (1990) Different mechanisms and multiple stages of LTP. Adv Exp Med Biol 268:359–368
Mattson MP, Keller JN, Begley JG (1998) Evidence for synaptic apoptosis. Exp Neurol 153:35–48
Mbadiwe T, Millis RM (2013) Epigenetics and autism. Autism Res Treat 2013:826156. doi:10.1155/2013/826156. (Epub 2013 Sep 15)
Mcallister AK (2007) Dynamic aspects of CNS synapse formation. Annu Rev Neurosci 30:450
Mcnair K, Davies CH, Cobb SR (2006) Plasticity-related regulation of the hippocampal proteome. Eur J Neurosci 23:575–580
Mehler MF (2008) Epigenetics and the nervous system. Annals Neurol 64:602–617
Mellios N, Sur M (2012) The Emerging Role of microRNAs in Schizophrenia and Autism Spectrum Disorders. Front Psychiatry 3:39 doi:10.3389/fpsyt.2012.00039. (eCollection@2012)
Merzenich MM, Kaas JH, Wall J, Nelson RJ, Sur M, Felleman D (1983) Topographic reorganization of somatosensory cortical areas 3b and 1 in adult monkeys following restricted deafferentation. Neuroscience 8:33–55
Merzenich MM, Nelson RJ, Stryker MP, Cynader MS, Schoppmann A, Zook JM (1984) Somatosensory cortical map changes following digit amputation in adult monkeys. J Comp Neurol 20(224):591–605
Minshew NJ, Keller TA (2010) The nature of brain dysfunction in autism: functional brain imaging studies. Curr Opin Neurol 23:124–130
Monnet-Tschudi F, Defaux A, Braissant O, Cagnon L, Zurich MG (2011) Methods to assess neuroinflammation. Curr.Protoc.Toxicol (Chap. 12:Unit12.19). doi:10.1002/0471140856.tx1219s50
Mor E, Cabilly Y, Goldshmit Y, Zalts H, Modai S, Edry L, Elroy-Stein O, Shomron N (2011) Species-specific microRNA roles elucidated following astrocyte activation. Nucleic Acids Res 39:3710–3723
Moretti P, Levenson JM, Battaglia F, Atkinson R, Teague R, Antalffy B, Armstrong D, Arancio O, Sweatt JD, Zoghbi HY (2006) Learning and memory and synaptic plasticity are impaired in a mouse model of Rett syndrome. J Neurosci 26:319–327
Morgan JT, Chana G, Pardo CA, Achim C, Semendeferi K, Buckwalter J, Courchesne E, Everall IP (2010) Microglial activation and increased microglial density observed in the dorsolateral prefrontal cortex in autism. Biol Psychiatry 68:368–376
Nedivi E, Hevroni D, Naot D, Israeli D, Citri Y (1993) Numerous candidate plasticity-related genes revealed by differential cDNA cloning. Nature 363:718–722
Nottebohm F (1989) From bird song to neurogenesis. Sci Am 260:74–79
Nottebohm F (2011) Song learning in birds offers a model for neuronal replacement in adult brain. In: Seki T et al (ed) Neurogenesis in the Adult Brain I: neurobiology. Springer Verlag. Berlin, pp 47–84
O'shea KS (2001) Directed differentiation of embryonic stem cells: genetic and epigenetic methods. Wound Rep Reg 9:443–459
Ohtsuka M, Inoko H, Kulski JK, Yoshimura S (2008) Major histocompatibility complex (Mhc) class Ib gene duplications, organization and expression patterns in mouse strain C57BL/6. BMC. Genomics 9:178. doi:10.1186/1471-2164-9-178
Olde Loohuis NF, Kos A, Martens GJ, Van BH, Nadif KN, Aschrafi A (2012) MicroRNA networks direct neuronal development and plasticity. Cell Mol Life Sci 69:89–102
Onore C, Careaga M, Ashwood P (2012a) The role of immune dysfunction in the pathophysiology of autism. Brain Behav Immun 26:383–392
Onore C, Van De Water J, Ashwood P (2012b) Decreased levels of EGF in plasma of children with autism spectrum disorder. Autism Res Treat 2012:205362. doi:10.1155/2012/205362. (Epub@2012 Mar 5)

Paluszkiewicz SM, Martin BS, Huntsman MM (2011) Fragile X Syndrome: the GABAergic system and circuit dysfunction. Dev Neurosci 33(5):349–364

Pardo CA, Vargas DL, Zimmerman AW (2005) Immunity, neuroglia and neuroinflammation in autism. Int Rev Psychiatry 17:485–495

Park CS, Tang SJ (2009) Regulation of microRNA expression by induction of bidirectional synaptic plasticity. J Mol Neurosci 38:50–56

Pate M, Damarla V, Chi DS, Negi S, Krishnaswamy G (2010) Endothelial cell biology: role in the inflammatory response. Adv Clin Chem 52:109–130

Peca J, Feng G (2012) Cellular and synaptic network defects in autism. Curr Opin Neurobiol 22:866–872

Pescosolido MF, Yang U, Sabbagh M, Morrow EM (2012) Lighting a path: genetic studies pinpoint neurodevelopmental mechanisms in autism and related disorders. Dialog Clin Neurosci 14:239–252

Pizzarelli R, Cherubini E (2011) Alterations of GABAergic signaling in autism spectrum disorders. Neural Plast 2011(2011):297153

Puckett RE, Lubin FD (2011) Epigenetic mechanisms in experience-driven memory formation and behavior. Epigenomics 3:649–664

Qureshi IA, Mehler MF (2010) Epigenetic mechanisms underlying human epileptic disorders and the process of epileptogenesis. Neurobiol Dis 39:53–60

Rinaldi T, Kulanagara K, Antoniello K, Markram H (2007) Elevated NMDA receptor levels and enhanced postsynaptic long-term potentiation induced by prenatal exposure to valproic acid. Proc Nat Acad Sci U S A 104:13501–13506

Rodriguez JI, Kern JK (2011) Evidence of microglial activation in autism and its possible role in brain underconnectivity. Neuron Glia Biol 7:205–213

Rosenzweig MR, Krech D, Bennett EL, Diamond MC (1962a) Effects of environmental complexity and training on brain chemistry and anatomy: a replication and extension. J Comp Physiol Psychol 55:429–437

Rosenzweig MR, Krech D, Bennett EL, Zolman JF (1962b) Variation in environmental complexity and brain measures. J Comp Physiol Psychol 55:1092–1095

Rossi F, Gianola S, Corvetti L (2007) Regulation of intrinsic neuronal properties for axon growth and regeneration. Prog Neurobiol 81:1–28

Sarachana T, Zhou R, Chen G, Manji HK, Hu VW (2010) Investigation of post-transcriptional gene regulatory networks associated with autism spectrum disorders by microRNA expression profiling of lymphoblastoid cell lines. Genome Med 2:23

Scheiffele P (2003) Cell-cell signalling during synapse formation in the CNS. Annu Rev Neurosci 26:485–508

Schratt GM, Tuebing F, Nigh EA, Kane CG, Sabatini ME, Kiebler M, Greenberg ME (2006) A brain-specific microRNA regulates dendritic spine development. Nature 439:283–289

Shatz CJ (2009) MHC class I: an unexpected role in neuronal plasticity. Neuron 64:40–45

Sinner J (2012) Defining synaesthesia. British J Psychol 103:1–15

Smith SD (2011) Approach to epigenetic analysis in language disorders. J Neurodev Disord 3:356–364

Song G, Ouyang G, Bao S (2005) The activation of Akt/PKB signaling pathway and cell survival. J Cell Mol Med 9:59–71

Streit WJ, Mrak RE, Griffin WS (2004) Microglia and neuroinflammation: a pathological perspective. J Neuroinflammation 1:14

Stubbs EG, Ritvo ER, Mason-Brothers A (1985) Autism and shared parental HLA antigens. J Am Acad Child Psychiatry 24:182–185

Stubbs G (1981) Shared parental HLA antigens and autism. Lancet 2:534

Syken J, Grandpre T, Kanold PO, Shatz CJ (2006) PirB restricts ocular-dominance plasticity in visual cortex. Science 313:1795–1800

Torres AR, Maciulis A, Stubbs EG, Cutler A, Odell D (2002) The transmission disequilibrium test suggests that HLA-DR4 and DR13 are linked to autism spectrum disorder. Hum Immunol 63:311–316

Torres AR, Sweeten TL, Cutler A, Bedke BJ, Fillmore M, Stubbs EG, Odell D (2006) The association and linkage of the HLA-A2 class I allele with autism. Hum Immunol 67:346–351
Tsankova N, Renthal W, Kumar A, Nestler EJ (2007) Epigenetic regulation in psychiatric disorders. Nat Rev Neurosci 8:355–367
Uddin LQ, Supekar K, Meno V (2013) Reconceptualizing functional brain connectivity in autism from a developmental perspective. Front Human Neurosci 7:1–11
Vaishnavi V, Manikandan M, Tiwary BK, Munirajan AK (2013) Insights on the functional impact of microRNAs present in autism-associated copy number variants. PLoS ONE 8:e56781
Valiente M, Marín O (2010) Neuronal migration mechanisms in development and disease. Curr Opin Neurobiol 20:68–78
Valko M, Leibfritz D, Moncol J, Cronin MT, Mazur M, Telser J (2007) Free radicals and antioxidants in normal physiological functions and human disease. Int J Biochem Cell Biol 39:44–84
Vargas DL, Nascimbene C, Krishnan C, Zimmerman AW, Pardo CA (2005) Neuroglial activation and neuroinflammation in the brain of patients with autism. Ann Neurol 57:67–81
Verly M, Verhoeven J, Zink I, Mantini D, Peeters R, Deprez S, Emsell L, Boets B, Noens I, Steyaert J, Lagae L, De CP, Rommel N, Sunaert S (2014a) Altered functional connectivity of the language network in ASD: role of classical language areas and cerebellum. Neuroimage Clin 4:374–82. doi:10.1016/j.nicl.2014.01.008. (eCollection@2014)
Verly M, Verhoeven J, Zink I, Mantini D, Van OL, Lagae L, Sunaert S, Rommel N (2014b) Structural and functional underconnectivity as a negative predictor for language in autism. Hum Brain Mapp 35:3602–3615
Volianskis A, Collingridge GL, Jensen MS (2013) The roles of STP and LTP in synaptic encoding. Peer J 1:e3. doi:10.7717/peerj.3. (Print@2013)
Waldman S, Nicholson S, Adilov N (2006) Does television causes autism. National Bureau of Economic Research. 22-8-2014
Wass S (2011) Distortions and disconnections: disrupted brain connectivity in autism. Brain Cogn 75:18–28
Webber AL, Wood J (2005) Amblyopia: prevalence, natural history, functional effects and treatment. Clin Exp Optom 88:365–375
Wright C, Turner JA, Calhoun VD, Perrone-Bizzozero N (2013) Potential impact of miR-137 and its targets in schizophrenia. Front Genet 4:58. doi:10.3389/fgene.2013.00058. (eCollection@2013)
Wu HH, Levitt P (2013) Prenatal expression of MET receptor tyrosine kinase in the fetal mouse dorsal raphe nuclei and the visceral motor/sensory brainstem. Dev Neurosci 35:1–16
Xu B, Hsu PK, Karayiorgou M, Gogos JA (2012) MicroRNA dysregulation in neuropsychiatric disorders and cognitive dysfunction. Neurobiol Dis 46:291–301
Zhang P, Lisman JE (2012) Activity-dependent regulation of synaptic strength by PSD-95 in CA1 neurons. J Neurophysiol 107:1058–1066
Ziats MN, Rennert OM (2013a) Aberrant expression of long noncoding RNAs in autistic brain. J Mol Neurosci 49:589–593
Ziats MN, Rennert OM (2013b) The cerebellum in autism: pathogenic or an anatomical beacon? Cerebellum 12:776–777
Zimmerman AW, Jyonouchi H, Comi AM, Connors SL, Milstien S, Varsou A, Heyes MP (2005) Cerebrospinal fluid and serum markers of inflammation in autism. Pediatr Neurol 33:195–201

Chapter 8
Neuroinflammation in Animal Models of Autism

Nadia Kazlauskas, Maria de los Angeles Robinson-Agramonte and Amaicha Mara Depino

Abstract In recent years, an increasing number of mouse and rat models of autism has been validated. These models have shown face validity, exhibiting alterations in behavior similar to those observed in autistic individuals. Moreover, some of these models were aimed to study different putative etiological factors of autism spectrum disorder (ASD), successfully showing construct validity. More recently, these animal models have been proved valuable to assess the contribution of neuroinflammation to ASD. On the one hand, eliciting inflammation in the central nervous system (CNS) during development was shown to result in behavioral alterations in mice similar to those observed in autistic individuals. On the other hand, neuroinflammation and malfunction of the immune system have been observed in different animal models of autism. The use of animal models has then contributed to stress the role and effects of neuroinflammation to the pathophysiology of autism.

Keywords Autism · Neuroinflammation · Animal models

8.1 Introduction

After a bodily injury or pathogen invasion, the immune system stereotyped response takes place. This mechanism of innate immunity is called inflammation, in contrast with the adaptive immune response, which is specific for each pathogen. This response is conducted by macrophages, dendritic cells, and leucocytes with the involvement of different kinds of cytokines and chemokines. The purpose is to isolate and eliminate the infectious agent, and to repair the damaged tissue.

A. M. Depino (✉) · N. Kazlauskas
Institute of Physiology, Molecular Biology and Neuroscience, CONICET—Department of Physiology, Molecular and Cellular Biology, Faculty of Exact and Natural Sciences, University of Buenos Aires, Int Guiraldes 2160, Buenos Aires, Argentina
e-mail: amaicha.depino@gmail.com

M. d. l. A. Robinson-Agramonte
Neuroimmunology Department, International Center for Neurological Restoration, Havana, Cuba

M. d. l. A. Robinson-Agramonte (ed.), *Translational Approaches to Autism Spectrum Disorder,* DOI 10.1007/978-3-319-16321-5_8

These mediators of peripheral inflammation are precluded from the brain by means of the blood brain barrier. This structure, formed by tight junctions of endothelial cells around the capillaries, separates the circulating blood from the brain's extracellular fluid and gives the central nervous system (CNS) an immune-privileged status.

The CNS has its own resident immune cells called microglia, a type of glial cell. These cells differentiate from hematopoietic progenitors in the bone marrow (BM) that travel to the brain, where they settle and complete their differentiation to microglia. There, they act as macrophages but are also involved in a variety of processes like synaptic pruning and cortical plasticity (Tremblay et al. 2011). In the case of nervous tissue damage, they go through morphological changes and transform from the ramified resting state to an activated amoeboid state with phagocytic activity (Kreutzberg 1996). During microglial activation, phagocytic microglia travel to the injury site, where they proliferate and secrete cytokines.

In addition to microglia, astrocytes are another type of glial cells. These are star-shaped cells with many cytoplasmatic processes that are involved in the physical structuring of the brain, and act as a skeleton during the development of the CNS to allow cellular migration. Besides this, their processes envelop neuronal synapses and regulate their activity through regulation of ion concentration in the extracellular space, growth factors production, and removal of excessive excitatory amino acids (Parpura et al. 2012). When the brain or the spinal cord is injured, astrocytes activate in a phenomenon called astrocytosis and participate in the removal of the damaged tissue, as well as in the scarring process. They also produce signals that can regulate the inflammatory response, such as cytokines, chemokines, and metalloproteases.

Neuroglial cells are important for the maintenance of neuronal functioning and homeostasis, and are involved in diverse phenomena such as cortical organization, neuroaxonal guidance, and synaptic plasticity (Fields and Stevens-Graham 2002). Since neuroglial cells are fulfilling roles both in the immune response and in neuronal functioning, we can conclude that in a neuroinflammatory state these cells can produce big neuronal and synaptic modifications that ultimately will affect behavior and cognitive processes.

On this line, there is growing evidence of neuroinflammatory and immunological abnormalities in patients with classical psychiatric disorders. In order to identify the agents that contribute to the etiology of these disorders, it is necessary to understand the neuronal mechanisms that regulate them. This is why the generation of animal models for the study of these diseases is essential.

8.2 Animal Models of Autism

Animal models of human diseases are valuable tools for studying the biological causes and mechanisms of disease, and for testing novel treatments. The use and validation of animal models of psychiatric diseases has emerged in the recent years and it has already proven valuable.

Homologous models of psychiatric diseases are not very common. So far, no animal has been shown to suffer from known psychiatric diseases, sharing its causes and symptoms, and responding similarly to the treatments validated in humans. So homologous models, which are frequently used for other diseases (e.g., obesity or cancer research), are not usually found in psychiatric biology.

A more limited purpose of animal models is to study the effects of novel treatments and to help developing new ones. This kind of model has been widely exploited for drugs targeting psychiatric symptoms. For example, rodents exposed to an open field experience a conflict between their fear-induced thigmotaxis and their desire to explore the novel environment (Belzung and Griebel 2001). A fearful animal would mainly explore the periphery of the field and avoid the center (especially if it is brightly illuminated). For years this test (the open-field test) has been used to test candidate anxiolytic drugs, proving a high predictive value: drugs that increased the exploration of the center of an open field in rodents showed an anxiolytic effect when administered to humans.

8.2.1 Criteria for Animal Models of Psychiatric Diseases

As in any experimental setup, two variables have to be taken into account for modeling a disease: the independent variable and the dependent variable. The independent variable is the modification introduced by the researcher; in this context, it is the intervention to an animal in order to induce an anomaly or specific behavior. The dependent variable is the result of that manipulation or the measurement performed.

In the case of animal models of human diseases and translational science, the main criterion these models should fulfill is to enable proper predictions. The animal model has *predictive validity* if a treatment tested on the animal has the same effects when it is provided to humans, as the example given for anxiolytic drugs. Sometimes it is possible to observe these effects without a clear understanding of the biological processes underlying them. For this reason, many researchers consider predictive models only partially valuable and prone to undesired side effects, warranted when the mechanisms involved in the disease and the treatment are not fully understood.

Other characteristics improve the value of animal models of psychiatric diseases. *Construct validity* warrants that the etiological cause of the disease is the same that is causing the phenotype in the animal. To evaluate the construct validity implies knowing the implicit or explicit hypothesis of the cause of the disease, for example, knowing how specific genes and their products can be associated to specific disorders. In this sense, numerous transgenic animals have been generated, with genetic mutations homologous to those identified in humans. The behavioral and pharmacological analyses of these models are important to identify the phenotypic changes that are associated with these mutations. They allow for comparing different hypotheses on the etiology of the disease and to explore novel treatments. They have also been valuable in testing the effects of different environmental factors.

Finally, animal models should have *face validity*: the alterations observed in the animal and the symptom observed in the human patient should be similar. Behind this request, the belief that the phenotype observed in the animal emerges from the same physiologic processes than the human disease is implicit, something that is not always true and in many cases not possible to test.

8.2.2 Validity of the Proposed Animal Models of Autism

So, for an animal model of autism, we can expect it to have construct, face, and predictive validity, or at least some of them. The more a model recapitulates the etiological factors of a disease, shares the phenotype/symptoms, and accurately predicts the outcome of proposed treatment, the more valuable the model would be to research in that psychiatric disease.

The strength of the current models of autism comes mainly from their face validity. Focus has been made to model the three core symptoms of autism, and different tests have been developed to measure these behaviors in rodents (Crawley 2004, 2007).

1. Reduced sociability. Mice being social mammals, reduced sociability can be readily detected and measured. Used tests include quantifying the interactions between two young or adult mice, measuring the social recognition and the response to a novel social stimulus, conditioned place preference, nesting patterns, or more complex analyses of the interactions between mice, such as gaze aversion (Balemans et al. 2010; de Theije et al. 2014; Defensor et al. 2011; DeLorey et al. 2008; Depino et al. 2011; Lucchina and Depino 2013; MacFabe et al. 2011; Malkova et al. 2012; McFarlane et al. 2008; Pearson et al. 2012; Peca et al. 2011; Schneider and Przewlocki 2005; Singer et al. 2009). In addition, the response to social stimuli (social odor) has also been measured (Hamilton et al. 2011). In general, showing reduced sociability is expected from a rodent model of autism.

2. Repetitive, stereotyped behaviors. Some rodent behaviors can be considered stereotyped when they are expressed but not needed. The amount of time spent in these behaviors can be measured or they can be quantified. Self-grooming, burrowing, and marble burying have been used to model this symptom (Amodeo et al. 2012; Depino et al. 2011; McFarlane et al. 2008; Pearson et al. 2011). Also, the reversal of a position habit in an appetitive T-maze task or in the Morris water maze has been used (de Theije et al. 2014; MacFabe et al. 2011; Malkova et al. 2012). Finally, the response to a novel object in comparison with a known object has been studied (Pearson et al. 2011).

3. Communication. Although some researchers have proposed a kind of language in mice and rats, most attempts to model the communication deficits observed in autistic individuals have focused on vocalizations and responses to vocalizations, parental retrieval of separated pups and the ultrasonic vocalization (USV) of the separated pups, and scent (urine) marking (Hamilton et al. 2011; Scattoni et al. 2008, 2011; Wohr et al. 2011).

Construct validity has been mainly searched using data from epidemiological studies to validate the way the model is generated. So, genetic models have focused on allelic variants observed in autistic patients (Abrahams and Geschwind 2008), and environmental models have chosen drugs proved to increase the incidence of ASD (e.g., valproic acid (VPA)) or other nongenetic risk factors (e.g., prenatal infection) to induce autism-related phenotypes in rodents.

Finally, predictive validity has been less explored in animal models of autism so far, when compared to other animal models of psychiatric diseases. In our opinion this could be due to the fact of this being a new field, and to the scarce availability of treatments for ASD patients. However, the relevance of these models on testing new treatments is being discussed in different reports (Meyer et al. 2005).

8.3 Studying the Role of Inflammation in Autism Using Animal Models

Two approaches can be envisioned to study the role of inflammation in autism. On the one hand, immune and inflammatory events can have an etiological role in autism. For these, animal models of different immune stimuli at different developmental ages can be developed and, so far, they have provided valuable information. On the other hand, the occurrence and magnitude of inflammation and immune alterations can be evaluated in animal models generated using other, noninflammatory environmental factors or genetic variants linked to autism.

In this section, we will review the animal models of autism generated through inflammatory stimuli or by activation of the immune system (Sect. 3.1). In addition, we will review the reports on immune alterations in noninflammatory and genetic models of autism (Sect. 3.2). This section does not aim to review all animal models of autism, but we will only report those in which inflammatory and immune alterations have been evaluated.

8.3.1 Inflammatory Models of Autism

Among the numerous environmental factors involved in the pathophysiology of ASD, maternal infection has been mentioned as the principal nongenetic cause of this disorder (Ciaranello and Ciaranello 1995). Epidemiological studies have shown a strong association between autism and maternal viral infection (Atladottir et al. 2010). In addition, numerous immune alterations have been reported in autism (see Chap. 9).

Taking these clinical evidences into account, some animal models have been generated by manipulating the immune system and evaluating the effect of an immune response on behaviors relevant to autism.

8.3.1.1 Maternal Immune Activation (MIA)

Numerous animal models have been yielded to study the effects that prenatal inflammation may generate on fetal neural development, as the maternal respiratory influenza infection (Shi et al. 2003; Fatemi et al. 2002), peridontal bacteria infection (Lin et al. 2003), or the MIA models (Smith et al. 2007). The last one is based on the injection of pregnant mice or rats with the synthetic double-stranded RNA polyinosinic:polycytidylic acid (polyI:C), a ligand for the Toll-like receptor 3 (TLR3), which is involved in the innate immune response to viral infection. The utilization of this generic inflammatory agent allows to model factors that are common to several infections and to become independent of the pathogen's nature.

Another factor to consider is the time when the infection occurs, as neither the maternal immune system functioning (Sargent 1993) nor the fetus developing nervous system susceptibility is homogeneous during pregnancy. In fact, an epidemiological study conducted in Denmark found an association between maternal viral infection in the first trimester or maternal bacterial infection in the second trimester, and ASD diagnosis in the child (Atladottir et al. 2010). Similar results have also been found in the MIA mouse model. Meyer et al. (2006) injected polyI:C in pregnant mice at gestation day 9 (GD9) or GD17; these times correspond to the transition from the first to the second trimester, and from the second to the third trimester of human pregnancy, respectively. They showed that behavioral alterations were dependent on the moment of immune challenge administration: GD9 but not GD17 prenatal immune activation reduced open-field exploration, suggesting an anxious phenotype, while MIA at GD17 but not at GD9 led to perseverative behavior, which has been implicated in autism, schizophrenia, and other neuropsychiatric disorders (Ridley 1994). They also found differences in the maternal serum and fetal brain cytokine profile responses after the mid and late prenatal immune activation.

A posterior study (Malkova et al. 2012) aimed to determine how capable of reproducing autistic symptoms the MIA model was. To this aim, they used several assays to evaluate autistic features on the offspring of pregnant dams injected with polyI:C or saline. They found that MIA pups produce fewer USVs and have an altered syllable repertoire. This deficit in communication is maintained until adulthood: adult male MIA offspring display reduced USV responses in the presence of both female and male social stimuli. These mice also present a deficit in sociability, as they spend less time exploring the social side in a three-chamber test. In addition, offspring of polyI:C injected dams display higher levels of repetitive and stereotyped behaviors (spent more time in self-grooming and buried a higher percentage of marbles).

In summary, this study showed that a model of prenatal exposure to an inflammatory stimulus could show the three core symptoms of ASD: impairments in communication, decreased social interaction, and repetitive and stereotyped behavior. Now, considering the immune alterations that are also observed in ASD patients, can a unique immune challenge event during the embryogenesis lead to a persistent immune deregulation in the adult offspring?

Hsiao et al. (2012) tackled this question by evaluating the profiles of several peripheral immune subtypes and the functionality of leukocyte lineages. Regulatory T cells (Tregs) participate in the innate and the adaptive responses as suppressor cells and have a reciprocal relationship with the pro-inflammatory IL-17-producing CD4+T cells (Th17). MIA offspring showed a systemic deficit in the number of Tregs and an elevated response to in vitro stimulation, given by higher levels of IL-6 and IL-17 secretion of CD4+T cells. They also found higher levels of Gr-1-positive cells, given by a skewed differentiation of the progenitor cells.

These abnormalities grant MIA mice a pro-inflammatory T-helper-cell phenotype that could be due to intrinsic hematopoietic stem cell (HSC) developmental programming. To test this hypothesis, Hsiao et al. transferred BM from polyI:C offspring into irradiated saline and polyI:C mice and found that the immune abnormalities were not transferred. This suggested a role for the HSC microenvironment in maintaining the pro-inflammatory phenotype seen in MIA mice.

Could these immunological alterations be responsible for the generation or for the maintenance of the ASD behavioral phenotype found in MIA mice? Or, are these just two independent unrelated manifestations of the disease? To assess these questions, they evaluated ASD-related behaviors in MIA offspring transplanted with saline BM. Interestingly, these mice failed to exhibit repetitive or anxious behaviors, but maintained the social preference deficit. These results add evidence to a group of studies that show the capacity of BM transplant on ameliorating symptoms in several neurological disorders (Chen et al. 2010; Kwan et al. 2012; Derecki et al. 2012), and also highlight the importance of peripheral environment in maintaining immune alterations in MIA mice.

8.3.1.2 Commensal Microbiota

It is known that microbiota can modulate host physiology, particularly immune functionality. But it was recently shown that it can also impact in mammalian brain development and behavior (Diaz Heijtz et al. 2011). Commensal bacteria affect social, emotional, and anxiety-like behaviors (Cryan and Dinan 2012; Collins et al. 2012).

ASD patients usually suffer from gastrointestinal (GI) distress, which also correlates with symptoms severity (Adams et al. 2011). Altered GI motility, increased intestinal permeability, and higher prevalence of inflammatory bowel disease are among other GI abnormalities (Kohane et al. 2012). As several studies have shown an atypical intestinal microbiota composition in ASD patients (Adams et al. 2011; Finegold et al. 2012; Kang et al. 2013; Parracho et al. 2005), a link between gut bacteria and GI abnormalities, and ASD behavioral phenotype has been proposed.

Hsiao et al. evaluated GI abnormalities in the MIA model of autism (Hsiao et al. 2013). They found that polyI:C offspring present a deficit in intestinal barrier integrity, altered intestinal cytokines profiles, and dysbiosis of gut microbiota. When MIA mice were treated with commensal bacteria, intestinal permeability and cytokine profiles were restored to normal levels. Moreover, this treatment ameliorated anxiety-like behavior, reduced levels of stereotyped marble burying, and restored communicative behavior; but it could not improve the sociability and social preference deficits.

These results indicate that there may be several neuronal circuits responsible for ASD behaviors with different sensitivity to gut microbiota modulation.

8.3.1.3 Propionic Acid

One possible mechanistic explanation for commensal bacteria modulating behavior is through the action of metabolic products in the CNS. Among the major molecules produced by enteric bacteria there is propionic acid (PPA), an intermediary in cellular fatty acid metabolism that is also present naturally in food and is often used as a food preservative.

Gut PPA can cross the gut–blood and blood–brain barriers, gaining access to the CNS, where it can alter neurotransmitter release and inhibit gap junctions through the acidification of cell interior (Bonnet et al. 2000; Karuri et al. 1993); also, elevated PPA levels stimulate the secretion of pro-inflammatory cytokines as IFN-gamma (Cavaglieri et al. 2003). These evidences indicate that PPA could potentially alter neuronal communication and behavior as well as modulate immune functioning. Moreover, as PPA levels increase during GI pathologies as those seen in ASD, this molecule may be a link between commensal bacteria and ASD phenotype.

Rats treated intracerebroventricularly (ICV) with PPA displayed restricted behavioral interest to a specific object and impaired social behavior (MacFabe et al. 2011; Shultz et al. 2008). In addition, reactive astrogliosis and activated microglia were observed in the hippocampus of PPA injected animals, suggesting an innate neuroinflammatory response (Shultz et al. 2008, 2009; MacFabe et al. 2011). These results support the utilization of PPA for the generation of a rat model of ASD.

8.3.1.4 Transforming Growth Factor-beta 1 (TGF-β1)

TGF-β1 is an anti-inflammatory cytokine involved in controlling immune responses, but it also has a role in brain development and glial function. There is evidence of altered levels of TGF-β1 in postmortem tissue of ASD individuals (Vargas et al. 2005). It can be hypothesized that adult ASD behavioral symptoms are a manifestation of an abnormal neuronal development given by altered levels of TGF-β1 at early life stages of development; or, alternatively, that this cytokine is important for regulating the expression of these behaviors in adulthood.

We studied TGF-β1 effects at two different scenarios by overexpressing this cytokine in the mouse hippocampus at different stages of development: PD14 or adulthood (Depino et al. 2011). We found that overexpression of TGF-β1 in the hippocampus of adult mice increased sociability and reduced self-grooming repetitive behavior, resulting in a phenotype opposed to what is expected for an ASD model. On the contrary, overexpression of TGF-β1 from PD14 led to reduced sociability and increased repetitive behavior, besides an increment in depression-related behavior.

These results showed that TGF-β1 is involved in the early postnatal programming of ASD-related behaviors, but it also has a role in modifying them in adulthood. This suggests that the increased levels of TGF-β1 observed in postmortem brain analyses of autistic patients could be a consequence of neuroinflammatory alterations during early neural development, resulting then in the behavioral phenotype. However, other cytokines besides TGF-β1 could be responsible for modulating and maintaining these symptoms in adulthood.

8.3.2 *Inflammation and Immune Alterations in Noninflammatory Environmental Models of Autism*

We have identified only one noninflammatory, environmental animal model of autism in which inflammation and immune function have been partially studied: the prenatal exposure to VPA. A number of studies have linked prenatal exposure to antiepileptic drugs, particularly VPA, to ASD (Bromley et al. 2008; Christensen et al. 2013; Moore et al. 2000; Williams et al. 2001). Based on these studies, rodent models of ASD have been generated.

A study identified that a single 400 mg/kg injection of VPA in pregnant rats at gestational day (GD) 12.5 results in low sociability of the offspring (Kim et al. 2011). Interestingly, the same study shows that this effect is not observed when dams are injected at GD7, 9.5 or 15, showing a critical window for the effect of VPA on social behavior. Lately, these results obtained in Sprague-Dawley rats were verified in other rat strains (Schneider et al. 2008; Schneider and Przewlocki 2005; Roullet et al. 2010; Bambini-Junior et al. 2011) and in different mice strains (Kataoka et al. 2012; Lucchina and Depino 2013; de Theije et al. 2013; Wagner et al. 2006; Yochum et al. 2008).

Recently, we reported long-lasting alterations in the inflammatory response in the VPA mouse model of autism (Lucchina and Depino 2013). Adult mice prenatally exposed to VPA showed peripheral and central exacerbated responses to a peripheral lipopolysaccharide (LPS) challenge. In the periphery, LPS induced an exacerbated activation of the hypothalamus–pituitary–adrenal (HPA) axis and an increased expression of the pro-inflammatory cytokine IL-6 in VPA mice compared with control mice. In the brain, VPA mice showed chronic microglia activation in the cerebellum, revealing a chronic gliosis. Moreover, peripheral LPS challenge augmented the number of microglial cells in the hippocampus and the expression of inflammatory cytokines in the cerebellum of VPA mice.

We found that the exacerbated response of the HPA axis to LPS is not observed early in life (Fig. 8.1), suggesting this is an effect reached after maturation of the nervous and immune systems. However, we found evidence of an increased number of microglial cells in the young cerebellum of VPA mice (Fig. 8.2), showing that the gliosis in this region is elicited early in life.

In addition, mice exposed prenatally to 500 mg/kg VPA showed intestinal inflammation at PD28 (de Theije et al. 2013). VPA mice showed loss of the epithelial

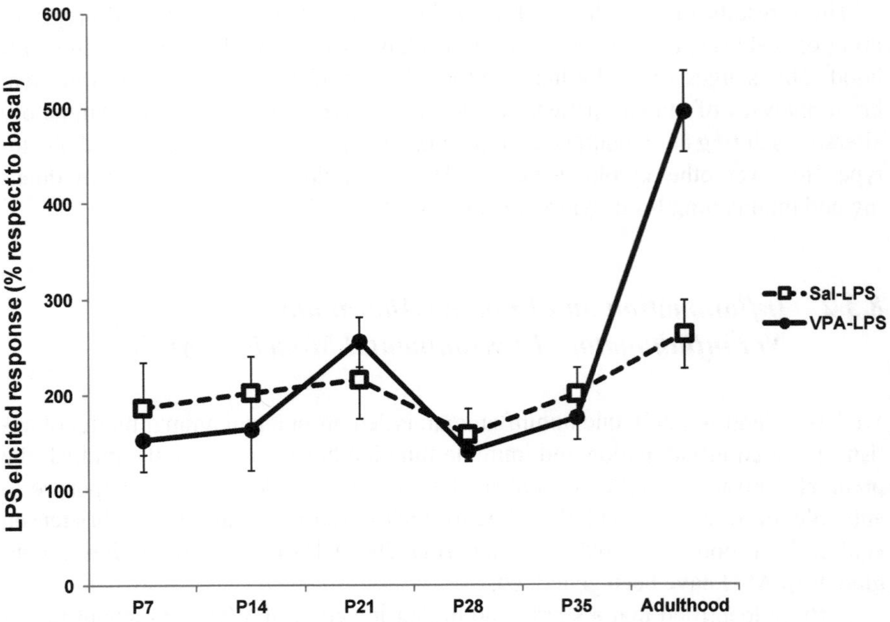

Fig. 8.1 Hypothalamus–pituitary–adrenal (*HPA*) axis response to peripheral LPS in mice prenatally exposed to VPA. Animals prenatally exposed to VPA show a normal HPA response postnatally, but an exacerbated response in adulthood. Plasma corticosterone levels were measured 2 h after a 25 mg/kg LPS i.p. injection. Values are expressed as a percentage with respect to animals prenatally exposed to saline and injected 2 h before with saline. *LPS* lipopolysaccharide, *VPA* valproic acid

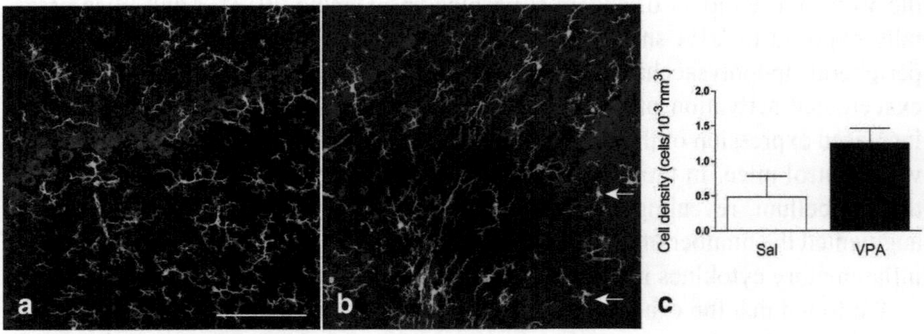

Fig. 8.2 Activated microglia in the postnatal cerebellar lobe 7 of VPA-exposed mice. Confocal microscopy photographies after IBA1 immunofluorescence of cerebellar tissue obtained at P21 from saline (**a**) and VPA (**b**) mice. The *arrows point* type II–III microglial cells, mainly observed in animals prenatally exposed to VPA. **c** Type II–III microglial cell density in the molecular layer of the cerebellar lobe 7. *VPA* valproic acid

barrier and neutrophilic infiltration in the ileum. These results give support to the hypothesis of an association between the GI diseases frequently observed in autistic individuals and their behavioral symptoms.

Although we still cannot distinguish between VPA causing the immune and behavioral alterations by two independent mechanisms or inflammation mediating the effects of VPA on behavior, the reports previously cited brought into light a valuable tool for these studies.

8.3.3 *Inflammation and Immune Alterations in Genetic Models of Autism*

Autism is a clinically heterogeneous disorder that in some cases can be associated with genetic disorders such as Rett and fragile X syndrome, or tuberous sclerosis. However, in the great majority of the patients a unique pathophysiological mechanism has not been able to be identified and the causes of the disorder remain unknown (Pardo and Eberhart 2007).

As diverse evidences suggest the existence of a heritable component in ASD, numerous animal models have been generated with the manipulation of candidate genes identified by means of population studies. Some mutant lines are based on monogenic aberrations and others on loci that confer susceptibility to the disease. Nevertheless, the attempts to link individual mutations with ASD have failed (Wassink et al. 2004), and nowadays there is consensus about the existence of a multiplicity of genes that confer susceptibility to the disease, which are influenced both by environmental factors and by their interactions with other genes.

In a study carried out by Moy et al. (2007), male mice from ten inbred strains were characterized for several autism-related behaviors such as sociability, preference for social novelty, and reversal learning of the spatial location of a reinforcer in two different mazes. Interestingly, they found that the BTBR T+tf/J strain displays reduced sociability and resistance to change in routine in the water maze. As these findings are consistent with an ASD phenotype, some groups have proposed to use the BTBR strain as a model of autism.

Other groups further characterized the autism-related behaviors in the BTBR strain (McFarlane et al. 2008; Amodeo et al. 2012; Pearson et al. 2011, 2012; Pobbe et al. 2010). They found that BTBR mice show reduced social approach, low reciprocal social interactions, impaired juvenile play, and absence of social conditioned place preference. Moreover, they display high levels of self-grooming and impaired social transmission of food preference as compared with C57BL/6J mice, and resistance to reversal learning. In addition, BTBR mice also show an unusual repertoire of USVs both as pups and as adults (Scattoni et al. 2008, 2011; Wohr et al. 2011).

These results show that BTBR mice display the three core features of autism: reduced sociability, repetitive behaviors, and impaired communication. The next step was to characterize the immune system of these mice, which was conducted by Heo et al. (2011). In this study, levels of brain-reactive antibodies were higher in

BTBR serum and brains, and they also presented elevated brain expression levels of pro-inflammatory cytokines such as IL-33, IL-18, and IL-1β. Additionally, BTBR mice presented higher susceptibility to listeriosis infection.

As the ability of macrophages to switch to an inflammatory profile is known to confer immunity to listeriosis, another study hypothesized an association between myeloid inflammation and ASD behavioral phenotype. Onore et al. (2013) aimed to examine the role of macrophage polarization (M1/M2) in the BTBR strain. BM macrophages from BTBR mice present higher inflammatory cytokine production with or without LPS stimulation. When M1 or M2 phenotypes were induced by the addition of polarizing cytokines to the media, BTBR macrophages exhibited a trend towards an M1 phenotype. They also evaluated the relationship between the behavioral and immune phenotypes by measuring social and repetitive behavior in each individual mouse and analyzing its correlation with the cytokine profile. The experiments showed that grooming behavior was associated with several cytokines, suggesting the existence of a direct relation between inflammatory cytokines and more impaired repetitive behavior on one hand, and of anti-inflammatory cytokines and behavioral improvement on the other. Conversely, they found very little association between social behavior and cytokine expression profiles, indicating that sociability may not be strongly affected by these cytokine alterations. This study suggests that the behavioral phenotype and the inflammatory immune profile of BTBR mice are not independent, but a connected phenomena.

8.3.4 Combining Genetic and Inflammatory Models

As it is now believed that ASD is not a consequence of a unique pathophysiological cause but the result of combining environmental, genetic, and epigenetic factors; it is important to study how these factors interact between themselves.

In line with this, Schwartzer et al. (2013) hypothesized that the genetic background of mice could influence the severity of the ASD phenotype elicited by an environmental risk factor such as the MIA. To test this hypothesis, BTBR and C57BL/6J (C57) pregnant dams were injected with polyI:C on GD12.5 and their offspring were tested for ASD-like behaviors. Both in BTBR mice as in C57 controls, maternal polyI:C injection altered USVs, impaired social behavior, and increased repetitive behaviors in the offspring. However, some of these effects were stronger on BTBR mice.

Interestingly, they also evaluated the induced response of splenocytes, measuring cytokine levels. Interestingly, they observed a significant strain x prenatal treatment interaction in the release of IL-17, with BTBR prenatally exposed to polyI:C secreting the highest amount of this cytokine.

These results are in line with evidences obtained in MIA and BTBR autism models, which demonstrate that both environmental factors and genetic predisposition can result in similar behavioral ASD phenotypes. Moreover, this study adds evidence to the synergistic hypothesis, as the combination of genetic and environmental factors exacerbates the ASD phenotype generated by either factor alone. Further studies would help elucidating the mechanisms through which inflammation can contribute to autism-related behaviors in both environmental and genetic models.

8.4 Conclusions

Animal models are very valuable tools to study the biological mechanisms underlying psychiatric disorders. Different groups have started using them for evaluating the role of immune alterations in the development of autism-related behaviors. Evidence shows that diverse immune stimuli can result in reduced sociability and increased repetitive behaviors. In addition, genetic and environmental models of autism showed immune dysfunction. Although the mechanisms are not elucidated, all these evidences suggest a role of immune factors in determining autism-related behavior and prove relevance to the studies with animal models of autism.

References

Abrahams BS, Geschwind DH (2008) Advances in autism genetics: on the threshold of a new neurobiology. Nat Rev Genet 9(5):341–355. doi:nrg2346 [pii] 10.1038/nrg2346

Adams JB, Johansen LJ, Powell LD, Quig D, Rubin RA (2011) Gastrointestinal flora and gastrointestinal status in children with autism–comparisons to typical children and correlation with autism severity. BMC Gastroenterol 11:22. doi:10.1186/1471-230X-11-22

Amodeo DA, Jones JH, Sweeney JA, Ragozzino ME (2012) Differences in BTBR T + tf/J and C57BL/6J mice on probabilistic reversal learning and stereotyped behaviors. Behav Brain Res 227(1):64–72. doi:10.1016/j.bbr.2011.10.032

Atladottir HO, Thorsen P, Ostergaard L, Schendel DE, Lemcke S, Abdallah M, Parner ET (2010) Maternal infection requiring hospitalization during pregnancy and autism spectrum disorders. J Autism Dev Disord 40(12):1423–1430. doi:10.1007/s10803-010-1006-y

Balemans MC, Huibers MM, Eikelenboom NW, Kuipers AJ, van Summeren RC, Pijpers MM, Tachibana M, Shinkai Y, van Bokhoven H, Van der Zee CE (2010) Reduced exploration, increased anxiety, and altered social behavior: Autistic-like features of euchromatin histone methyltransferase 1 heterozygous knockout mice. Behav Brain Res 208(1):47–55. doi:10.1016/j.bbr.2009.11.008

Bambini-Junior V, Rodrigues L, Behr GA, Moreira JC, Riesgo R, Gottfried C (2011) Animal model of autism induced by prenatal exposure to valproate: behavioral changes and liver parameters. Brain Res 1408:8–16. doi:10.1016/j.brainres.2011.06.015

Belzung C, Griebel G (2001) Measuring normal and pathological anxiety-like behaviour in mice: a review. Behav Brain Res 125(1–2):141–149

Bonnet U, Bingmann D, Wiemann M (2000) Intracellular pH modulates spontaneous and epileptiform bioelectric activity of hippocampal CA3-neurones. Eur Neuropsychopharmacol 10(2):97–103

Bromley RL, Mawer G, Clayton-Smith J, Baker GA (2008) Autism spectrum disorders following in utero exposure to antiepileptic drugs. Neurology 71(23):1923–1924. doi:71/23/1923 [pii] 10.1212/01.wnl.0000339399.64213.1a

Cavaglieri CR, Nishiyama A, Fernandes LC, Curi R, Miles EA, Calder PC (2003) Differential effects of short-chain fatty acids on proliferation and production of pro- and anti-inflammatory cytokines by cultured lymphocytes. Life Sci 73(13):1683–1690

Chen SK, Tvrdik P, Peden E, Cho S, Wu S, Spangrude G, Capecchi MR (2010) Hematopoietic origin of pathological grooming in Hoxb8 mutant mice. Cell 141(5):775–785. doi:S0092-8674(10)00374-0 [pii] 10.1016/j.cell.2010.03.055

Christensen J, Gronborg TK, Sorensen MJ, Schendel D, Parner ET, Pedersen LH, Vestergaard M (2013) Prenatal valproate exposure and risk of autism spectrum disorders and childhood autism. JAMA 309(16):1696–1703. doi:10.1001/jama.2013.2270

Ciaranello AL, Ciaranello RD (1995) The neurobiology of infantile autism. Annu Rev Neurosci 18:101–128. doi:10.1146/annurev.ne.18.030195.000533

Collins SM, Surette M, Bercik P (2012) The interplay between the intestinal microbiota and the brain. Nat Rev Microbiol 10(11):735–742. doi:10.1038/nrmicro2876

Crawley JN (2004) Designing mouse behavioral tasks relevant to autistic-like behaviors. Ment Retard Dev Disabil Res Rev 10(4):248–258. doi:10.1002/mrdd.20039

Crawley JN (2007) Mouse behavioral assays relevant to the symptoms of autism. Brain Pathol 17(4):448–459. doi:BPA096 [pii] 10.1111/j.1750-3639.2007.00096.x

Cryan JF, Dinan TG (2012) Mind-altering microorganisms: the impact of the gut microbiota on brain and behaviour. Nat Rev Neurosci 13(10):701–712. doi:10.1038/nrn3346

de Theije CG, Koelink PJ, Korte-Bouws GA, Lopes da Silva S, Korte SM, Olivier B, Garssen J, Kraneveld AD (2013) Intestinal inflammation in a murine model of autism spectrum disorders. Brain Behav Immun. doi:10.1016/j.bbi.2013.12.004

de Theije CG, Wu J, Koelink PJ, Korte-Bouws GA, Borre Y, Kas MJ, Lopes da Silva S, Korte SM, Olivier B, Garssen J, Kraneveld AD (2014) Autistic-like behavioural and neurochemical changes in a mouse model of food allergy. Behav Brain Res 261:265–274. doi:10.1016/j.bbr.2013.12.008

Defensor EB, Pearson BL, Pobbe RL, Bolivar VJ, Blanchard DC, Blanchard RJ (2011) A novel social proximity test suggests patterns of social avoidance and gaze aversion-like behavior in BTBR T + tf/J mice. Behav Brain Res 217(2):302–308. doi:S0166-4328(10)00718-7 [pii] 10.1016/j.bbr.2010.10.033

DeLorey TM, Sahbaie P, Hashemi E, Homanics GE, Clark JD (2008) Gabrb3 gene deficient mice exhibit impaired social and exploratory behaviors, deficits in non-selective attention and hypoplasia of cerebellar vermal lobules: a potential model of autism spectrum disorder. Behav Brain Res 187(2):207–220. doi:S0166-4328(07)00483-4 [pii] 10.1016/j.bbr.2007.09.009

Depino AM, Lucchina L, Pitossi F (2011) Early and adult hippocampal TGF-beta1 overexpression have opposite effects on behavior. Brain Behav Immun 25(8):1582–1591. doi:S0889-1591(11)00180-2 [pii] 10.1016/j.bbi.2011.05.007

Derecki NC, Cronk JC, Lu Z, Xu E, Abbott SB, Guyenet PG, Kipnis J (2012) Wild-type microglia arrest pathology in a mouse model of Rett syndrome. Nature 484(7392):105–109. doi:nature10907 [pii] 10.1038/nature10907

Diaz Heijtz R, Wang S, Anuar F, Qian Y, Bjorkholm B, Samuelsson A, Hibberd ML, Forssberg H, Pettersson S (2011) Normal gut microbiota modulates brain development and behavior. Proc Natl Acad Sci U S A 108(7):3047–3052. doi:10.1073/pnas.1010529108

Fatemi SH, Earle J, Kanodia R, Kist D, Emamian ES, Patterson PH, Shi L, Sidwell R (2002) Prenatal viral infection leads to pyramidal cell atrophy and macrocephaly in adulthood: implications for genesis of autism and schizophrenia. Cell Mol Neurobiol 22(1):25–33

Finegold SM, Downes J, Summanen PH (2012) Microbiology of regressive autism. Anaerobe 18(2):260–262. doi:10.1016/j.anaerobe.2011.12.018

Hamilton SM, Spencer CM, Harrison WR, Yuva-Paylor LA, Graham DF, Daza RA, Hevner RF, Overbeek PA, Paylor R (2011) Multiple autism-like behaviors in a novel transgenic mouse model. Behav Brain Res 218(1):29–41. doi:S0166-4328(10)00751-5 [pii] 10.1016/j.bbr.2010.11.026

Heo Y, Zhang Y, Gao D, Miller VM, Lawrence DA (2011) Aberrant immune responses in a mouse with behavioral disorders. PLoS ONE 6(7):e20912. doi:10.1371/journal.pone.0020912 PONE-D-11-02222 [pii]

Hsiao EY, McBride SW, Chow J, Mazmanian SK, Patterson PH (2012) Modeling an autism risk factor in mice leads to permanent immune dysregulation. Proc Natl Acad Sci U S A 109(31):12776–12781. doi:1202556109 [pii] 10.1073/pnas.1202556109

Hsiao EY, McBride SW, Hsien S, Sharon G, Hyde ER, McCue T, Codelli JA, Chow J, Reisman SE, Petrosino JF, Patterson PH, Mazmanian SK (2013) Microbiota modulate behavioral and physiological abnormalities associated with neurodevelopmental disorders. Cell 155(7):1451–1463. doi:S0092-8674(13)01473-6 [pii] 10.1016/j.cell.2013.11.024

Kang DW, Park JG, Ilhan ZE, Wallstrom G, Labaer J, Adams JB, Krajmalnik-Brown R (2013) Reduced incidence of Prevotella and other fermenters in intestinal microflora of autistic children. PLoS ONE 8(7):e68322. doi:10.1371/journal.pone.0068322

Karuri AR, Dobrowsky E, Tannock IF (1993) Selective cellular acidification and toxicity of weak organic acids in an acidic microenvironment. Br J Cancer 68(6):1080–1087

Kataoka S, Takuma K, Hara Y, Maeda Y, Ago Y, Matsuda T (2012) Autism-like behaviours with transient histone hyperacetylation in mice treated prenatally with valproic acid. Int J Neuropsychopharmacol:1–13. doi:S1461145711001714 [pii] 10.1017/S1461145711001714

Kim KC, Kim P, Go HS, Choi CS, Yang SI, Cheong JH, Shin CY, Ko KH (2011) The critical period of valproate exposure to induce autistic symptoms in Sprague-Dawley rats. Toxicol Lett 201(2):137–142. doi:S0378-4274(10)01813-8 [pii] 10.1016/j.toxlet.2010.12.018

Kohane IS, McMurry A, Weber G, MacFadden D, Rappaport L, Kunkel L, Bickel J, Wattanasin N, Spence S, Murphy S, Churchill S (2012) The co-morbidity burden of children and young adults with autism spectrum disorders. PLoS ONE 7(4):e33224. doi:10.1371/journal.pone.0033224 PONE-D-12-00026 [pii]

Kreutzberg GW (1996) Microglia: a sensor for pathological events in the CNS. Trends Neurosci 19(8):312–318. doi:0166223696100497 [pii]

Kwan W, Magnusson A, Chou A, Adame A, Carson MJ, Kohsaka S, Masliah E, Moller T, Ransohoff R, Tabrizi SJ, Bjorkqvist M, Muchowski PJ (2012) Bone marrow transplantation confers modest benefits in mouse models of Huntington's disease. J Neurosci 32(1):133–142. doi:10.1523/JNEUROSCI.4846-11.2012

Lin TY, Kao HT, Hsieh SH, Huang YC, Chiu CH, Chou YH, Yang PH, Lin RI, Tsao KC, Hsu KH, Chang LY (2003) Neonatal enterovirus infections: emphasis on risk factors of severe and fatal infections. Pediatr Infect Dis J 22(10):889–894. doi:10.1097/01.inf.0000091294.63706.f3

Lucchina L, Depino AM (2013) Altered peripheral and central inflammatory responses in a mouse model of autism. Autism Res. doi:10.1002/aur.1338

MacFabe DF, Cain NE, Boon F, Ossenkopp KP, Cain DP (2011) Effects of the enteric bacterial metabolic product propionic acid on object-directed behavior, social behavior, cognition, and neuroinflammation in adolescent rats: relevance to autism spectrum disorder. Behav Brain Res 217(1):47–54. doi:S0166-4328(10)00677-7 [pii] 10.1016/j.bbr.2010.10.005

Malkova NV, Yu CZ, Hsiao EY, Moore MJ, Patterson PH (2012) Maternal immune activation yields offspring displaying mouse versions of the three core symptoms of autism. Brain Behav Immun 26(4):607–616. doi:S0889-1591(12)00013-X [pii] 10.1016/j.bbi.2012.01.011

McFarlane HG, Kusek GK, Yang M, Phoenix JL, Bolivar VJ, Crawley JN (2008) Autism-like behavioral phenotypes in BTBR T + tf/J mice. Genes Brain Behav 7(2):152–163. doi:GBB330 [pii] 10.1111/j.1601-183X.2007.00330.x

Meyer U, Feldon J, Schedlowski M, Yee BK (2005) Towards an immuno-precipitated neurodevelopmental animal model of schizophrenia. Neurosci Biobehav Rev 29(6):913–947

Meyer U, Nyffeler M, Engler A, Urwyler A, Schedlowski M, Knuesel I, Yee BK, Feldon J (2006) The time of prenatal immune challenge determines the specificity of inflammation-mediated brain and behavioral pathology. J Neurosci 26(18):4752–4762

Moore SJ, Turnpenny P, Quinn A, Glover S, Lloyd DJ, Montgomery T, Dean JC (2000) A clinical study of 57 children with fetal anticonvulsant syndromes. J Med Genet 37(7):489–497

Moy SS, Nadler JJ, Young NB, Perez A, Holloway LP, Barbaro RP, Barbaro JR, Wilson LM, Threadgill DW, Lauder JM, Magnuson TR, Crawley JN (2007) Mouse behavioral tasks relevant to autism: phenotypes of 10 inbred strains. Behav Brain Res 176(1):4–20

Onore CE, Careaga M, Babineau BA, Schwartzer JJ, Berman RF, Ashwood P (2013) Inflammatory macrophage phenotype in BTBR T + tf/J mice. Front Neurosci 7:158. doi:10.3389/fnins.2013.00158

Pardo CA, Eberhart CG (2007) The neurobiology of autism. Brain Pathol 17(4):434–447

Parpura V, Heneka MT, Montana V, Oliet SH, Schousboe A, Haydon PG, Stout RF Jr, Spray DC, Reichenbach A, Pannicke T, Pekny M, Pekna M, Zorec R, Verkhratsky A (2012) Glial cells in (patho)physiology. J Neurochem 121(1):4–27. doi:10.1111/j.1471-4159.2012.07664.x

Parracho HM, Bingham MO, Gibson GR, McCartney AL (2005) Differences between the gut microflora of children with autistic spectrum disorders and that of healthy children. J Med Microbiol 54(Pt 10):987–991. doi:54/10/987 [pii] 10.1099/jmm.0.46101-0

Pearson BL, Pobbe RL, Defensor EB, Oasay L, Bolivar VJ, Blanchard DC, Blanchard RJ (2011) Motor and cognitive stereotypies in the BTBR T + tf/J mouse model of autism. Genes Brain Behav 10(2):228–235. doi:10.1111/j.1601-183X.2010.00659.x

Pearson BL, Bettis JK, Meyza KZ, Yamamoto LY, Blanchard DC, Blanchard RJ (2012) Absence of social conditioned place preference in BTBR T + tf/J mice: relevance for social motivation testing in rodent models of autism. Behav Brain Res 233(1):99–104. doi:10.1016/j.bbr.2012.04.040

Peca J, Feliciano C, Ting JT, Wang W, Wells MF, Venkatraman TN, Lascola CD, Fu Z, Feng G (2011) Shank3 mutant mice display autistic-like behaviours and striatal dysfunction. Nature 472(7344):437–442. doi:10.1038/nature09965

Pobbe RL, Pearson BL, Defensor EB, Bolivar VJ, Blanchard DC, Blanchard RJ (2010) Expression of social behaviors of C57BL/6J versus BTBR inbred mouse strains in the visible burrow system. Behav Brain Res 214(2):443–449. doi:10.1016/j.bbr.2010.06.025

Ridley RM (1994) The psychology of perserverative and stereotyped behaviour. Prog Neurobiol 44(2):221–231

Roullet FI, Wollaston L, Decatanzaro D, Foster JA (2010) Behavioral and molecular changes in the mouse in response to prenatal exposure to the anti-epileptic drug valproic acid. Neuroscience 170(2):514–522. doi:S0306-4522(10)00941-3 [pii] 10.1016/j.neuroscience.2010.06.069

Sargent IL (1993) Maternal and fetal immune responses during pregnancy. Exp Clin Immunogenet 10(2):85–102

Scattoni ML, Gandhy SU, Ricceri L, Crawley JN (2008) Unusual repertoire of vocalizations in the BTBR T + tf/J mouse model of autism. PLoS ONE 3(8):e3067. doi:10.1371/journal.pone.0003067

Scattoni ML, Ricceri L, Crawley JN (2011) Unusual repertoire of vocalizations in adult BTBR T + tf/J mice during three types of social encounters. Genes Brain Behav 10(1):44–56. doi:10.1111/j.1601-183X.2010.00623.x

Schneider T, Przewlocki R (2005) Behavioral alterations in rats prenatally exposed to valproic acid: animal model of autism. Neuropsychopharmacology 30(1):80–89. doi:10.1038/sj.npp.1300518 1300518 [pii]

Schneider T, Roman A, Basta-Kaim A, Kubera M, Budziszewska B, Schneider K, Przewlocki R (2008) Gender-specific behavioral and immunological alterations in an animal model of autism induced by prenatal exposure to valproic acid. Psychoneuroendocrinology 33(6):728–740. doi:S0306-4530(08)00061-9 [pii] 10.1016/j.psyneuen.2008.02.011

Schwartzer JJ, Careaga M, Onore CE, Rushakoff JA, Berman RF, Ashwood P (2013) Maternal immune activation and strain specific interactions in the development of autism-like behaviors in mice. Transl Psychiatry 3:e240. doi:10.1038/tp.2013.16

Shi L, Fatemi SH, Sidwell RW, Patterson PH (2003) Maternal influenza infection causes marked behavioral and pharmacological changes in the offspring. J Neurosci 23(1):297–302

Shultz SR, MacFabe DF, Ossenkopp KP, Scratch S, Whelan J, Taylor R, Cain DP (2008) Intracerebroventricular injection of propionic acid, an enteric bacterial metabolic end-product, impairs social behavior in the rat: implications for an animal model of autism. Neuropharmacology 54(6):901–911. doi:S0028-3908(08)00035-X [pii] 10.1016/j.neuropharm.2008.01.013

Shultz SR, Macfabe DF, Martin S, Jackson J, Taylor R, Boon F, Ossenkopp KP, Cain DP (2009) Intracerebroventricular injections of the enteric bacterial metabolic product propionic acid impair cognition and sensorimotor ability in the Long-Evans rat: further development of a rodent model of autism. Behav Brain Res 200(1):33–41. doi:S0166-4328(08)00727-4 [pii] 10.1016/j.bbr.2008.12.023

Singer HS, Morris C, Gause C, Pollard M, Zimmerman AW, Pletnikov M (2009) Prenatal exposure to antibodies from mothers of children with autism produces neurobehavioral alterations: a pregnant dam mouse model. J Neuroimmunol 211(1–2):39–48. doi:S0165-5728(09)00105-2 [pii] 10.1016/j.jneuroim.2009.03.011

Smith SE, Li J, Garbett K, Mirnics K, Patterson PH (2007) Maternal immune activation alters fetal brain development through interleukin-6. J Neurosci 27(40):10695–10702. doi:27/40/10695 [pii] 10.1523/JNEUROSCI.2178-07.2007

Tremblay ME, Stevens B, Sierra A, Wake H, Bessis A, Nimmerjahn A (2011) The role of microglia in the healthy brain. J Neurosci 31(45):16064–16069. doi:10.1523/JNEUROSCI.4158-11.2011

Vargas DL, Nascimbene C, Krishnan C, Zimmerman AW, Pardo CA (2005) Neuroglial activation and neuroinflammation in the brain of patients with autism. Ann Neurol 57(1):67–81

Wagner GC, Reuhl KR, Cheh M, McRae P, Halladay AK (2006) A new neurobehavioral model of autism in mice: pre- and postnatal exposure to sodium valproate. J Autism Dev Disord 36(6):779–793. doi:10.1007/s10803-006-0117-y

Wassink TH, Brzustowicz LM, Bartlett CW, Szatmari P (2004) The search for autism disease genes. Ment Retard Dev Disabil Res Rev 10(4):272–283. doi:10.1002/mrdd.20041

Williams G, King J, Cunningham M, Stephan M, Kerr B, Hersh JH (2001) Fetal valproate syndrome and autism: additional evidence of an association. Dev Med Child Neurol 43(3):202–206

Wohr M, Roullet FI, Crawley JN (2011) Reduced scent marking and ultrasonic vocalizations in the BTBR T + tf/J mouse model of autism. Genes Brain Behav 10(1):35–43. doi:10.1111/j.1601-183X.2010.00582.x

Yochum CL, Dowling P, Reuhl KR, Wagner GC, Ming X (2008) VPA-induced apoptosis and behavioral deficits in neonatal mice. Brain Res 1203:126–132. doi:S0006-8993(08)00096-6 [pii] 10.1016/j.brainres.2008.01.055

Chapter 9
Understanding on Neuroimmunology in Autism Spectrum Disorder

Amaicha Mara Depino and Maria de los Angeles Robinson-Agramonte

Abstract Autism spectrum disorders (ASD) are a group of severe pervasive neurodevelopmental disorders (PDD) affecting between 1 and 3% of pediatric populations. It is characterized by a highly variable impairment in social interaction and verbal and nonverbal communication, as well as by a stereotyped repetitive behavior, learning problems, and aloofness. ASD behavioral symptoms are frequently accompanied by immunological derangements, including cellular immune dysregulation, chronic inflammatory states, and neuroimmune alterations. Currently, the involvement of the immune pathology in autism remains unclear, and we consider that a better understanding would be useful for earlier clinical and therapeutic interventions. The main aim of this chapter is to review the most current aspects regarding the etiology of autism, with particular reference to the contribution of inflammatory events occurring in the periphery and into the brain, and how they can influence the abnormal development of the offspring and modulate the typical behaviors frequently observed in autism.

Keywords Neuroimmunology · ASD · Neurodevelopmental disorders · Cytokine · T cells · mTOR pathway · Innate immunity

9.1 Introduction

Autism spectrum disorders (ASD) are a group of severe pervasive neurodevelopmental disorders (PDD) present in early childhood, which affect between 1 and 3% of the pediatric population and are characterized by a highly variable combination of impairment in social interaction, deficits in verbal and nonverbal communication and restricted, repetitive, and stereotyped patterns of behavior, interests, and activities (Chakrabarti and

M. d. l. A. Robinson-Agramonte (✉)
Neuroimmunology Department, International Center for Neurological Restoration, Havana, Cuba
e-mail: maria.robinson@infomed.sld.cu

A. M. Depino
Institute of Physiology, Molecular Biology and Neuroscience, CONICET—Department of Physiology, Molecular and Cellular Biology, Faculty of Exact and Natural Sciences, University of Buenos Aires, Int Guiraldes 2160, Buenos Aires, Argentina

M. d. l. A. Robinson-Agramonte (ed.), *Translational Approaches to Autism Spectrum Disorder*, DOI 10.1007/978-3-319-16321-5_9

Fombonne 2005; Fombonne 2005). With a four times higher incidence in boys than in girls, autism is accompanied by core symptoms of neurobehavioral and immunological derangements, including aberrant sensitivity to sensory stimulation, anxiety, and decreased cellular immune capacity (American Psychiatric Association 1994). This chapter is aimed to describe current aspects of immune and other epigenetic factors in autism, as well as to establish possible bridges between those interacting within and outside the brain, looking for a major understanding of the disease.

9.2 Immune Factors Contributing to the Etiology of Autism

Various environmental factors have been linked to an increased incidence of ASD (reviewed in (Depino 2013)). In particular, immunological triggers, such as prenatal, perinatal, and early postnatal infections, have been associated with autism (Niehus and Lord 2006; Rosen et al. 2007; Vojdani et al. 2003; Ashwood and Wakefield 2006). Evidence includes increased rates of infection in neonates who were later diagnosed with autism (Rosen et al. 2007), incidence of congenital rubella, cytomegalovirus, and other viral infections in ASD (Chess 1971; Stubbs et al. 1984), as well as several other immune abnormalities associated with increased risk for ASD (Ashwood et al. 2006; Cohly and Panja 2005), which are referred to as consequences of autoimmune phenomena around pregnancy.

An aberrant immune function has been reported in individuals affected by ASD for nearly 40 years and among these findings, there is a long-standing focus in the autoimmune processes (Comi et al. 1999) that could be playing a pathogenic role, at least in a subgroup of patients with autism, following either the autoimmune reaction generated by the global immune deregulation during gestation or by the abnormal interaction between maternal or fetal environmental and genetic factors, occurring at prenatal or perinatal periods (Fombonne 2005; Hertz-Picciotto et al. 2006).

Maternal and/or prenatal immunological events can affect neural development by different mechanisms. On the one hand, they can elicit an imbalance toward proinflammatory cytokines that can in turn affect both cytokine expression within the brain and the neural development (Merrill 1992). For example, an increase in the levels of the chemokine MCP-1 have been reported in the amniotic fluid of children later diagnosed with ASD (Abdallah et al. 2012), as well as the occurrence of autoimmune pathologies during pregnancy can result in the maternal transfer of autoantibodies from mother to child.

9.3 Functional Plasticity of Immune Cells and ASD

The immune system consists of innate and adaptive immunity. Although both innate and adaptive immunity are crucial for the organisms response to infection, these responses can be harmful and lead to autoimmune diseases (Medzhitov and

Janeway 2000). Abnormal innate and adaptive immune responses have been reported in ASD.

9.3.1 Innate Immunity in ASD

Monocytes, the key regulators of immune responses, express different variants of Toll-like receptors (TLR). The TLR are evolutionarily conserved to recognize pathogen-associated molecular patterns (PAMPs) in bacteria, viruses, fungi, and parasites (Abrahams et al. 2006). They form the major family of pattern recognition receptors involved in innate immunity and monocyte responses. When compared with cells of typically developing children, monocyte cultures from children with ASD showed increased secretion of proinflammatory IL-1β, IL-6, and TNFα cytokines after stimulation with TLR2 ligands. In addition, the IL-1β secretion induced by TLR4 was also increased, and a decrease in IL-1β, IL-6, GM-CSF, and TNFα responses after TLR9 stimulation was observed (Enstrom et al. 2010a). These results indicate that monocyte cultures from children with ASD are more sensitive to respond to TLRs signals. Hence, it is possible to consider that dysfunction in the response of these cells could result in long-term immune changes leading to a potential adverse neuroimmune interaction in autism.

Caspases are key regulators of apoptosis and their activity is essential for several cellular processes, including differentiation, activation, and nuclear reprogramming pathways (Kersse et al. 2007; Lamkanfi et al. 2007). Indeed, caspases are highly involved in the T lymphocytes proliferation and the activation and differentiation of monocytes (Wallach et al. 2010; Lamkanfi et al. 2007). The peripheral blood mononuclear cells (PBMCs) from children with autism may also show a significant upregulation of the caspase pathway, following published results for Siniscalco et al. (2012).

9.3.2 Adaptive Immunity in ASD

Adaptive immunity is mediated by T and B lymphocytes. T helper cells constitute the key factors to establish an effective immune response. Under the differential influence of cytokines and other specific stimuli, naïve T helper cells differentiate into the four functional well-known types: Th1, Th2, T-regulator (Treg), and Th17. These cell types have been linked to ASD (Fig. 9.1).

Th1 cells are induced by IL-12 and they express T-bet, a Th1 transcription factor. They produce IFN-γ as their main proinflammatory cytokine, which is responsible for some pathogenic effects observed in both autoimmune and inflammatory disorders (Romagnani et al. 2000; Romagnani 2000). Th2 helper cells are induced by IL-4 and they express GATA3 (Mjosberg et al. 2012; Hoyler et al. 2012). This cell type produces IL-5, IL-9, IL-10, and IL-13 cytokines, all associated with the humoral immunity.

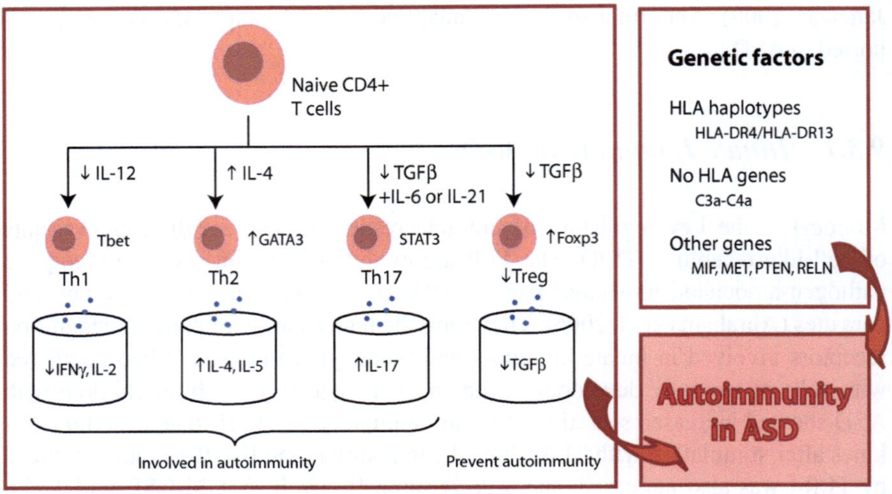

Fig. 9.1 Functional plasticity of T cells and associated genetic factors may mediate autoimmunity in autism. Naïve CD4+ T cells become differentiated into four different effector cell subtypes, which show functional activities promoting or preventing autoimmunity. Reported altered levels of specific cell types, transcription factors, and secreted cytokines are shown. The increase in Th2 function and decrease in Th1 and Treg function, promote an immune activation leading to autoimmunity. In addition, variants of genes associated with autoimmunity have been identified in ASD, particularly haplotypes of the human leukocyte antigen (HLA-DR4/HLA-DR13) and mutations in no HLA genes related to innate immunity (C3a–C4a genes) and signaling pathways like PTEN

The balance of Th1/Th2 cells is tightly controlled, and an overactivity of either of these cell types results in immune disease. There is evidence of a shift from Th1 to Th2 cells in autism, observed as decreased production of IL-2 and IFN-γ and an increase in IL-4 levels (van Gent et al. 1997; Gupta et al. 1998), Fig. 9.1. However, an increase of IL-12 levels in the plasma of autistic patients has been reported (Singh 1996). Authors proposed that this abnormal production of IL-12 could be a compensatory mechanism in ASD individuals, but this hypothesis has not been further tested. Moreover, a more recent report showed that cells from ASD patients express more Th1 and Th2 cytokines in absence of a compensatory increase of IL-10, a cytokine that modulates the inflammatory response (Molloy et al. 2006), reinforcing again the immune deregulation observed in these children.

The expression of GATA3, a transcription factor that is needed for Th2 differentiation, has been associated with autism. In a neuron-like PC-12 cell model, GATA3 expression levels are increased after their treatment with valproate, thalidomide, or alcohol, three teratogens known to cause ASD (Rout and Clausen 2009). As GATA3 plays a key role in the differentiation of some neuronal types (like serotonergic neurons; (van Doorninck et al. 1999)), it has been hypothesized that an excessive differentiation of neurons expressing GATA3 could cause anomalies in the development of brain resulting in autistic symptoms. So, GATA3 could represent a molecular link between immune and neuronal dysfunction in ASD.

Naturally occurring regulatory T (nTreg) cells are a central component of peripheral tolerance as they suppress the adaptive T cell response, show an anti-inflammatory role, and prevent autoimmunity. These cells were originally defined by high CD25 expression and primarily characterized by the expression of the transcription factor Foxp3 (forkhead/winged helix transcription factor). After differentiation, they produce TGF-β. Stable expression of Foxp3 is essential for the development, homeostasis, and suppressive function of nTreg cells (Sun et al. 2007; Liu et al. 2010; Ueda et al. 2012). At least one report has shown a decrease in the number of Treg cells in children with autism (Mostafa et al. 2010), Fig. 9.1. In addition, plasma levels of TGF-β were overall decreased in autistic individuals and they showed an inverse correlation with behavioral scores (Ashwood et al. 2008b; Okada et al. 2007), indicating that regulatory T cell responses are decreased in ASD and that the lack of suppressor activity of the immune system could be involved in the expression of autistic behavior.

Th17 cells, a newly defined lineage of T helper cells, differ from other T helper cells subsets both in their gene expression and regulatory function (Ueda et al. 2012; Dong 2008). These cells are remarked by the production of the hallmark cytokines IL-17 and IL-17F. The IL-17 cytokine family is composed by 6 cytokines, IL-17 A–F, and besides its role for Th17 differentiation, they mediate inflammatory functions and autoimmunity. Stimulation of T cells by TGF-β together with IL-6 or IL-21 promotes Th17 development (Harrington et al. 2006). IL-17 is augmented in autistic children (Suzuki et al. 2011), Fig. 9.1.

The nature of Foxp3 expression in induced Treg (iTreg) cells in vitro (Ueda et al. 2012) combined with the potential of Foxp3+ nTreg cells to differentiate into Th17 cells in the presence of IL-6 and TGF-β, suggested that Foxp3-expressing Treg cell subsets manifest functional adaptation under certain conditions (Ayyoub et al. 2012; Ueda et al. 2012; Bettelli et al. 2007; Maloy 2008). Although there is not a complete understanding on the signaling mechanism inducing Th17 differentiation, studies have shown that the key transcription factors in determining the differentiation of this T cell lineage are retinoid-related orphan receptor γt (RORγt) and RORα which can be induced by the combination of IL-6 and TGF-β (Ueda et al. 2012; Yurchenko et al. 2012). Thymic or peripheral Foxp3+ nTreg cells may also manifest prominent functional plasticity and readily reprogram into Th1 and Th17 effectors' cells, particularly in the gut microenvironment or sites of parasitic infection (Ahern et al. 2008). As it will be discussed later, this could lead to the chronic gastrointestinal inflammation observed in ASD.

9.4 Other Genes Regulating the Immune System in ASD

Family studies have provided some evidences for a higher genetic risk for autism in siblings compared to the general population (Holt et al. 2010). Early findings suggested that mothers of children with ASD share HLA haplotypes with their children more often than the typically developing mother–child pairs. In addition, these children showed a higher rate of these haplotypes when compared with the healthy population (Gregg et al. 2008). HLA haplotypes, considered among the strongest predictors of risk for autoimmune conditions, have also been associated with ASD (Fig. 9.1).

Studies in children and adults with autism have indicated other various candidate genes, many of them related to the immune system. These genes include the null allele of C4B in class III region, MHC haplotypes B44-SC30-DR4, HLA-DR β1, DR4, and DR13, all reinforcing also the factors lined as guttering in the development of autoimmunity (Daniels et al. 1995; Lee et al. 2006; Torres et al. 2006; Warren et al. 1995). Genes regulating signal pathways related to the development of both neural and immune systems have also been referred to as altered in ASD: macrophage migration inhibitory factor (MIF) (Grigorenko et al. 2008), MET encoding tyrosine kinase, (Correll et al. 2004), the serine and threonine kinase C gene PRKCB1 (Lintas et al. 2009), PTEN (Herman et al. 2007; Butler et al. 2005), and the reelin gene RELN (Garbett et al. 2008; Pardo and Eberhart 2007; Fig. 9.1).

9.5 Maternal Immune Activation and the Risk for ASD

As mentioned above, not only the adaptive but also the innate immune system response appears altered in ASD. These abnormal responses in children with autism could result from prenatal infection or exposure to toxic substances, and/or they could have a genetic component.

Various viral and bacterial infections during pregnancy have been linked to a higher risk for autism. More than 40 years ago, a higher risk for autism was reported in the offspring of maternal rubella infection cases (Chess 1971). Since then, prenatal exposure to different virus have been linked to increased risk for ASD, including herpes simplex virus (DeLong et al. 1981; Ghaziuddin et al. 1992; Gillberg 1986; Greer et al. 1989), cytomegalovirus (Ivarsson et al. 1990; Markowitz 1983; Stubbs 1978; Yamashita et al. 2003), measles (Ring et al. 1997; Singh and Jensen 2003), and viral meningitis (Barak et al. 1999; Ring et al. 1997). Moreover, a larger study showed an increase in the prevalence of autism in children born to mothers hospitalized for viral infections during the first trimester of pregnancy or for bacterial infections during the second trimester (Atladottir et al. 2010). These evidences support the idea that the maternal response to viral infection can be a risk factor for autism (Ciaranello and Ciaranello 1995).

How can different pathogenic infections, all result in an increased risk for autism? Evidence from experimental animal models show that prenatal and early life inflammation can alter behavior later in life (Coe and Lubach 2005; Meyer et al. 2006, 2007, 2009; Depino 2006; see also Chap. 8). Therefore, different infectious agents, autoimmune diseases, and stressful events could elicit the expression of immune factors affecting brain development and resulting in the behavioral symptoms (Fig. 9.2).

Interestingly, fetal inflammatory response syndrome (FIRS) is a condition in which, despite an absence of cultivable microorganisms, neonates have very high circulating levels of inflammatory cytokines, such as IL-1, IL-6, IL-8, and TNF-α

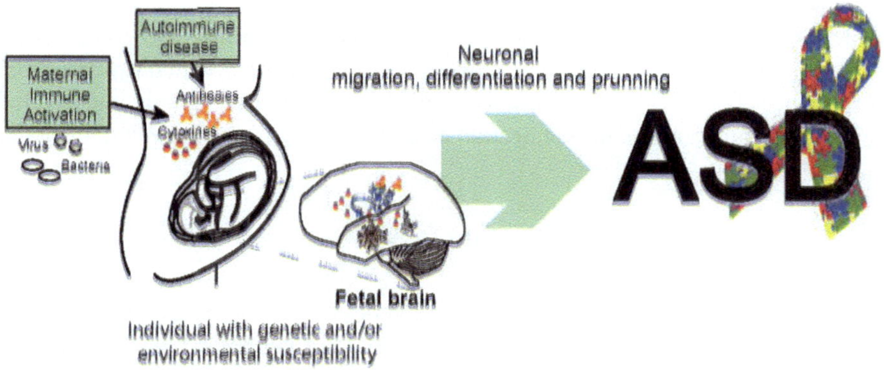

Fig. 9.2 Maternal immune activation can contribute to the development of ASD. Upon an infection, maternal cytokines and antibodies can reach the embryo's brain and affect its development. In individuals genetically susceptible, these effects on neuronal development could result in ASD

(Madsen-Bouterse et al. 2010). Studies in animal models have shown that viral infection of the placenta triggers a fetal inflammatory response similar to the one observed in FIRS, even though the virus is not able to reach the fetus. In the case of human FIRS, these cytokines have been shown to affect the CNS and the circulatory system (Deverman and Patterson 2009). Fetal morphological abnormalities in the animals have also been reported, probably caused by fetal proinflammatory cytokines such as IL-1, TNF-α, MCP-1, MIP-1β, and IFN-γ, and it has been suggested that beyond morphological effects on the fetal brain, the presence of FIRS increases the risk of autism, schizophrenia, neurosensorial deficits, and psychosis. Interestingly, all these factors induced at the neonatal period can predispose to diseases in adulthood (Romero et al. 2007). Due to this, we propose that an inflammatory response in the placenta, altering the cytokine balance in the fetus, may affect the normal development of the fetal immune system leading to anomalous responses during childhood or later in life. In all these cases the parasite may never reach the placenta, but the inflammatory process at this level may affect the normal fetal development.

9.6 Gastrointestinal Immune Pathology Increase Susceptibility to ASD

Recently, it has been hypothesized that the presence of gastrointestinal inflammation makes a child with a genetic predisposition for ASD more prone to express the autistic phenotype, or that it increases the severity of the autistic behavior. A recent meta-analysis found that the reported prevalence of gastrointestinal symptoms in individuals with ASD can range from 9 to 91 % (Buie et al. 2010; Coury et al. 2012). The variability observed could be mainly due to the lack of appropriate (nonrelated) control subjects, but the data is however consistent with a high prevalence of gastrointestinal symptoms and disorders in ASD individuals.

Gastrointestinal symptoms include abdominal pain or discomfort, constipation, diarrhea, and bloating (Afzal et al. 2003; Horvath and Perman 2002; Buie et al. 2010). In addition, autism severity shows a strong correlation with gastrointestinal disturbances (Adams et al. 2011), since studies in the field have mainly investigated immune dysfunction and the connection between the gut microbiome and the brain.

9.6.1 Gut Immune Dysfunction in ASD

Mucosal immune cells represent around 70% of an individual's immune cells, so their malfunction can affect the gastrointestinal environment. Clinical and pathological studies in ASD show an apparently atypical gastrointestinal immunopathology characterized by chronic ileocolonic lymphoid nodular hyperplasia and enterocolitis as key features, accompanied by a variable degree of acute inflammation and eosinophil infiltration (Wakefield et al. 2000; Furlano et al. 2001; Torrente et al. 2002). Increased intestinal permeability in ASD also suggests an ongoing inflammatory process that may perturb the intestinal barrier function in these children (D'Eufemia et al. 1996), probably through the release of peptides and mediators by the enteric nervous system, which can induce activated mast cells to release neurotoxins, proinflammatory mediators, and vasoactive substances that getting into the circulation can cross the blood–brain barrier (BBB) and affect the neuronal functions (Sun et al. 2007), Fig. 9.3.

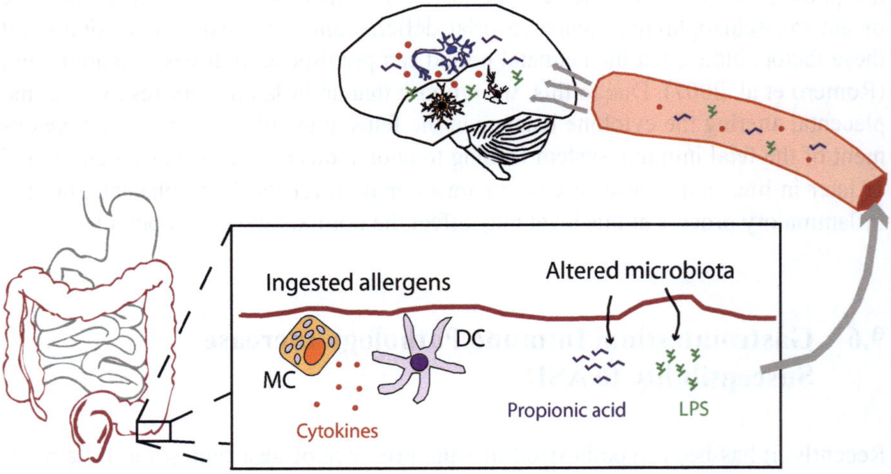

Fig. 9.3 Disturbances in the autistic gastrointestinal tract can affect brain function. Ingested allergens can elicit the secretion of cytokines (IL-1β, IL-4, IL-5, IL-6, IL-12, IL-13, IFN-γ, TNF-α) by mast cells (*MC*) and dendritic cells (*DC*). These cytokines can reach the brain through the circulation and cross the blood–brain barrier (BBB) or stimulate the brain endothelial cells, to induce a local immune response and cause a dysregulated neuronal homeostasis. In addition, changes in the microbiota can result in high levels of LPS and propionic acid reaching the circulation and acting in the brain, affecting brain function

In this context, a hypothesis relevant to understanding some alterations of mucosal immunity in autism could involve the functional plasticity of peripheral Foxp3+ nTreg cells (Fig. 9.4). Foxp3+ nTreg cells can become into Th1 and Th17 effectors cells, particularly in the gut microenvironment or sites of parasitic infection, following inflammatory signals that modulate the signal pathway and influence the stability of the Foxp3+ nTreg cell phenotype (Ahern et al. 2008; Liu et al. 2010). Interestingly, it has been shown that the T cell receptor controls Foxp3 expression via the PI3K-Akt-mTor signaling pathway (Sauer et al. 2008). The mammalian target of rapamycin (mTor) is also involved in synaptic function and plasticity (Hoeffer and

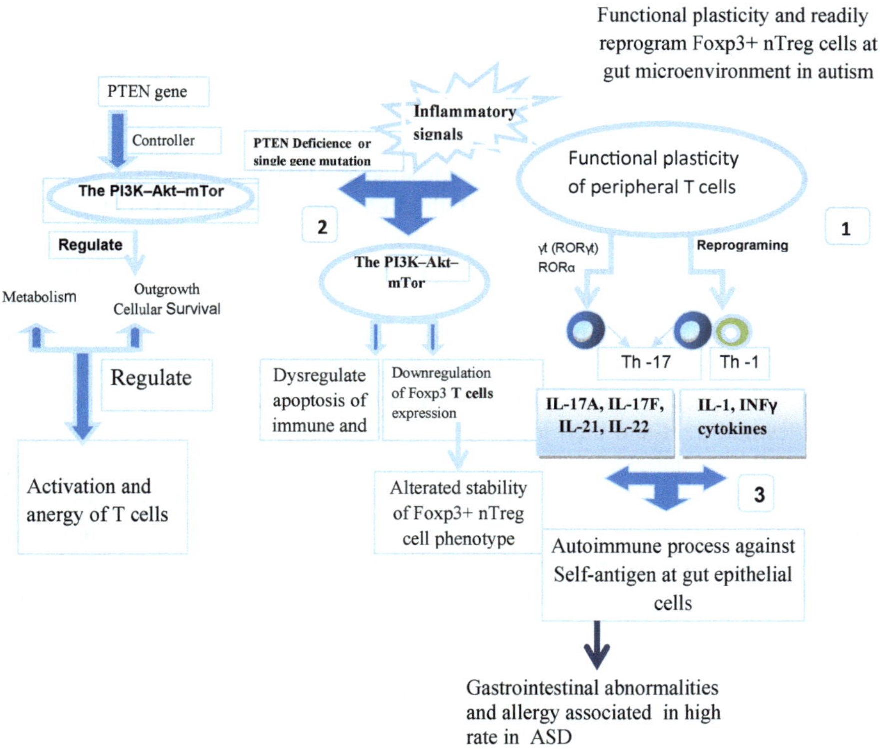

Fig. 9.4 Prominent functional plasticity and readily reprogram Foxp3+ nTreg cells. A hypothesis relevant to understand some alterations of mucosal immunity in autism could involve the functional plasticity of peripheral Foxp3+ nTreg cells. (*1*) Foxp3+ nTreg cells can convert into Th1 and Th17 effector cells particularly in the gut microenvironment, after an infection or other inflammatory signals. (*2*) PTEN gene is a controller of mTor pathway. Inflammatory signals and/or gene mutation are also able to modulate mTor pathway influencing the stability of the Foxp3+ nTreg cell phenotype, altered in high rate in ASD. So, it has been seen that mTor inhibition stabilizes Foxp3 expression by suppression of nTreg cells. (*3*) Aberrant immune mechanism at the gastrointestinal tract, followed by the influence of ASD-associated cytokines and lymphocytes induce an increased permeability of the blood–brain barrier (BBB). Activated lymphocytes and serum cytokines crossing BBB, bind to brain endothelial cells to induce an immune response into the brain with a dysregulated neuronal homeostasis

Klann 2010). Therefore, dysfunction in this pathway could be responsible for the aberrant cellular immune response (Mondino and Mueller 2007) and the changes in both neuronal and immune cell survival, both reported in autism. In addition, this reinforces the suggestion that cell survival mechanisms are altered in autism (Fatemi et al. 2001; Engstrom et al. 2003). Moreover, mutations in the phosphatase and tensin homolog (PTEN), a major upstream regulator of the mTOR pathway, have been found in association with autism (Rodriguez-Escudero et al. 2011). These findings could provide new directions toward a better understanding of the immune dysfunction in autism, as well as the identification of new therapeutic targets for ASD in the future.

Alternatively, co-localization of immunoglobulin and complement components on the epithelial membrane, in both stomach and duodenal specimens, could be indicative of an autoimmune process against self-antigen contained within epithelial cells and an inflammatory degradation that could contribute to disruption of the intestinal barrier function (Torrente et al. 2002; Furlano et al. 2001).

Gastrointestinal pathology in ASD has been also evidenced by a higher number of helper and cytotoxic T lymphocyte infiltration and CD19-positive B cells at duodenum, ileum, and colon in biopsies of children with ASD (Ashwood et al. 2003; Torrente et al. 2002), compared to histologically noninflamed controls, been more enhanced at terminal ileum and colon level (Ashwood et al. 2003; Torrente et al. 2002; Furlano et al. 2001; de Theije et al. 2011). Thus, some pathways of intestinal inflammation may be susceptible to establish a neuroimmune interaction linked to this disturbance in support of this hypothesis in autism. Dietary antigens and bacteria, as well as their products, can become a target for immune regulation at the gastrointestinal tract and then influence the nervous system activity. So, ingested antigens entering the gut mucosa can access the Peyer's patch or can be taken up by antigen presenting cell (APCs) that come to the Peyer's patch or mesenteric lymph node, and there they interact with naïve T lymphocytes to initiate an adaptive immune response. In the context of chronic inflammation, memory T and B cells are activated, damaging the epithelial cell layer, releasing cytokines, and thereby increasing the intestinal permeability and causing a gastrointestinal inflammation similar to what has been found in autism (de Magistris et al. 2010; D'Eufemia et al. 1996). In addition, it has been demonstrated that dietary derived antigen peptides are able to bind to the lymphocyte receptors and tissue enzymes CD69 and CD26, respectively, resulting in autoimmune reaction in children with autism (Vojdani et al. 2003; Siniscalco and Antonucci 2013).

In this respect, mast cell–neuron interactions occurring in the gastrointestinal tract might influence behavior via mast cells or other immune cells, which are able to trigger enteric neurons to convey information through afferent pathways in vagal and spinal nerves to the CNS (Adams et al. 2011).

Another pathway to reinforce the neuroimmune interaction surrounding the intestinal inflammation in autism is related to serotonergic neurotransmission. Serotonin is both a neurotransmitter and a mediator of inflammation, implicated early in the pathogenesis of autism (Schain and Freedman 1961). Hyperserotonemia in one third of the patients with ASD (Hanley et al. 1977) and a decrease in the capacity of 5-HT synthesis in the brain of these children (Chugani et al. 1999) have been demonstrated in autism (Anderson et al. 1987; Minderaa et al. 1987; Hoekstra et al. 2010). ASD-related hyperserotonemia was proposed to be from genetic (Holt et al. 2010; Sousa et al. 2010; Coutinho et al. 2007), gastrointestinal (Anderson et al. 1987; Hoekstra et al. 2010; Holt et al. 2010; Hoekstra and Wheelwright 2009; Nijmeijer et al. 2010), or immune origin (Jyonouchi et al. 2005; Burgess et al. 2006; Warren et al. 1986).

For example, an increased utilization of dietary tryptophan by the gut argue less tryptophan available for passage through the BBB with a reduced 5-HT levels in the brain leading to the mood and cognitive dysfunctions reported in ASD (Cappiello et al. 1996; McDougle et al. 1996). Nevertheless, more research is required to establish from this analysis a therapeutic target in ASD, either by providing dietary tryptophan or by pharmaceutical treatments based on selective serotonin reuptake inhibitors.

Finally, during gastrointestinal inflammation, the immune cells produce cytokines and chemokines in the gut, which are released into the intravascular compartment to pass all organs, including the brain (Banks 2012; Erickson et al. 2012). Therefore, gastrointestinal inflammation may influence the brain homeostasis and module characteristic behavior pattern in these patients.

9.6.2 Gut Microbiome and ASD

A probable factor contributing to the gastrointestinal disturbances in autistic individuals is the observed abnormal composition of gut microbiota (Finegold et al. 2002; Parracho et al. 2005; Adams et al. 2011). Fecal microflora is qualitatively different in autistic subjects compared with controls and siblings (Finegold et al. 2010; Kang et al. 2013), and they present an imbalance in beneficial bacteria (Adams et al. 2011). Specifically, the flora of autistic children is rich in *Bacteroidetes* and the genus *Desulfovibrio* (Finegold et al. 2010; Finegold et al. 2012).

How can the microflora of autistic children contribute to their behavioral symptoms? Bacteria can produce a variety of harmful substances (Fig. 9.3). For example, *Bacteroidetes* produce propionic acid as end product of their metabolism. The intracerebroventricular injection of propionic acid in rats causes a decrease in sociability, an increase in repetitive behaviors and gliosis, and they have been proposed as a rat model of ASD (MacFabe et al. 2011). The genus *Bacteroidetes* is highly abundant in the fecal material of autistic children (Finegold et al. 2010; Kang et al. 2013).

Bacteria of the genus *Desulfovibrio* can be found in about 50 % of autistic subjects, in some siblings of autistic children, but not in normal controls (Finegold 2011). *Desulfovibrio* is a Gram-negative anaerobic vibrio that produces lipopolysaccharide. Serum endotoxin levels are significantly higher in autistic children than in control subjects, and they show an inverse correlation with socialization in patients with severe autism (Emanuele et al. 2010).

Finally, gut microbiota have an important role in the development and function of both the innate and the adaptive immune systems (Round et al. 2010; Olszak et al. 2012). So, altered microbiota composition could also argue the immune alterations observed in ASD individuals.

In a small open-label trial with vancomycin (an antibiotic that is not absorbed in the gut), children with regressive autism showed behavioral improvement (Sandler et al. 2000). However, the use of probiotics or antibiotic in the treatment of autism needs to be further evaluated.

9.7 Cytokines as Mediators for Autism Development

Autoimmune diseases, viral infection, gastrointestinal disturbances, and other proposed etiological and pathophysiological pathways in ASD share a common result: Changes in the levels of cytokines produced either in the periphery and/or within the brain. Hence, cytokines could be considered as the main molecular effectors acting on the brain to alter development and behavior.

A neuroimmune hypothesis on the role of cytokines in autism can be that: (1) proinflammatory cytokines arising from maternal inflammation, infection or allergy during pregnancy, cross the placenta, enter the fetal circulation and pass through the fetal BBB; (2) an excess of proinflammatory cytokines released into the fetal brain cause aberrant neurogenesis, alters synaptogenesis and neuronal function; and (3) altered brain development results in the behavioral symptoms, including reduced sociability, stereotypic and repetitive behavior, hyperactivity, attention deficit, anxiety, and hypersensitivity to environmental stimuli.

After an immune response is elicited, mechanisms are triggered to limit this response either in time or in magnitude. When these mechanisms fail, autoimmune or other pathological results can derive and affect the organism. So, as important to mount an effective immune response is to end it when the stimulus is eliminated. To this purpose, anti-inflammatory cytokines and endocrine factors are a key to limit the immune response, and their failure can lead to long-lasting immune diseases.

The analysis of cytokine profiles in autism pathology reinforces the implication of an immune deregulation in this disorder. The plasma of autistic patients presents increased levels of proinflammatory cytokines such as IL-1β, IL-6, and TNF-α (Ashwood et al. 2011a; Malik et al. 2011), and reduced levels of anti-inflammatory cytokines such as TGF-β1 (Ashwood et al. 2008a). In addition, peripheral blood mononuclear cells (PBMC) from subjects with ASD show greater lipopolysaccharide-stimulated production of IL-1β (Enstrom et al. 2010b), TNF-α, and IL-6 (Jyonouchi et al. 2008) than those from controls. Moreover, PBMC from ASD children stimulated with phytohemagglutinin have increased activation of both Th2 and Th1 cytokines (Molloy et al. 2006). Finally, a multiplex assay for cytokines and chemokines in plasma from individuals with high-functioning ASD and matched controls showed increased plasma concentrations of IL-1β, IL-1RA, IL-5, IL-8, IL-12(p70), IL-13, IL-17, and GRO-α (Suzuki et al. 2011). On the other hand, Onore and colleagues showed normal levels of IL-17 and decreased IL-23 in autistic children (Onore et al. 2009). Another study revealed significant differences to IL-1β, IL-12(p70), IL-12(p40), and IL-17 in relationship to electrophysiological events, identified at EEG register and focused at temporal region (Robinson et al. unpublished results).

Altered brain cytokine levels have also been reported in ASD patients. Postmortem studies have shown high levels of TGF-β, IL-6, and MCP-1, as well as activated microglia in the brain of ASD subjects compared to controls (Vargas et al. 2005). Moreover, increased expression of cytokines has also been reported in cerebrospinal fluid of ASD patients (Chez et al. 2007; Garbett et al. 2008; Li et al. 2009).

Peripheral cytokines can not only alter immune functions but they can also affect different behaviors (Konsman et al. 2002). Cytokines can reach the brain directly, or through activation of different communication pathways, including the gut–brain and immune–brain pathways.

During an allergic process, immunoglobulins and neuropeptides (e.g., substance P, NGF, VIP, and neurotensin) activate mastoid cells and basophils to release cytokines, which can in turn trigger enteric neurons. As the mast cell–neurons interaction occurs at the gut level, an allergic reaction at this level might influence behavior following the action of enteric neurons, which are triggered by mast cell to convey information through the vagal and spinal afferent pathways into the CNS. A hypothesis emphasizing the effect of allergy-linked inflammatory mediators in autism underlines that molecules like IL-6 and histamine promote gene transcription of estradiol and retinoic acid. This results in an overexposure of the fetus to retinoic acid and estradiol, which has been described as causing brain abnormalities similar to those observed in ASD, such as cerebellar malformations, abnormalities of dopaminergic system, and cranial nerve abnormalities (Garay et al. 2012). Moreover, fetal overexposure to estradiol can result in anxiety, hyperactivity and attention deficit, stereotype and repetitive movement, and motor deficits (de Theije et al. 2011).

For centuries, the structured system of the BBB, the lack of conventional lymphatic drainage, and the low response to alloantigen inoculated in the CNS, were arguments to consider this compartment as a privileged site. Nevertheless, today it is well known that there is no such privilege, but rather the main differences of the immune response in the brain, regarding the periphery, are in the kinetics and the degree of regulation of the different stages of the immune response at this level, which will take place under the functional control of the immune response and influence of cellular mediators acting in a regulated quantitative and temporally balanced sequence. Moreover, numerous cytokines are able to cross the BBB, including IL-1β, IL-6, and TNF-α (McLay et al. 2000; Prat et al. 2001), whose plasma levels are enhanced in autistic patients (Ashwood et al. 2011b; Malik et al. 2011). After reaching the brain, specific cytokines can affect brain development and function. For example, TGF-β involved in diverse aspects of development (Ashwood et al. 2008b; Letterio and Roberts 1998). Levels of TGF-β have been found to be reduced in the plasma of autistic children and adults, in correlation with more severe behavioral scores (Ashwood et al. 2008b), while higher levels of TGF-β were described in postmortem brain and cerebrospinal fluid samples in ASD (Vargas et al. 2005). This apparent contradiction could result from different roles of this cytokine in the brain and in the periphery, and/or result from different temporal roles of this cytokine during development. Indeed, using an animal model, it was shown that the overexpression of TGF-β1 in the brain can have opposite effects on behavior, dependant on the age of the animals (Depino et al. 2011).

Macrophage inhibitory factor (MIF), other proinflammatory immune regulator that is constitutively expressed in the brain tissue, has an important influence on neural and endocrine systems (Grigorenko et al. 2008). Recently, high plasma levels of MIF were observed in ASD individuals with severe behavioral symptoms, and genotyping studies linked MIF to ASD and showed the presence of two polymorphisms in its promoter region (Grigorenko et al. 2008).

In addition to others functions in the immune system, TNF-αα and IL-1β regulate brain processes such as neuron plasticity, long-term potentiation (LTP) and associative learning, and memory (more information in Chap. 7) (Depino et al. 2004).

In the brain, cytokines can also act on neuroglial cells to induce neuroinflammation. Several factors released by astrocytes might be important for the induction and maintenance of the BBB, as manifested by the appearance of endothelial tight junctions in cells unsheathed by astrocytic endfeet. In the healthy CNS, astrocytes and microglia play important roles in neuronal function and homeostasis, as they are both fundamentally involved in cortical organization, neuroaxonal guidance, and synaptic transmission (Bilbo and Schwarz 2009).

Astrocytes and microglia, produce neurotrophic factors, cytokines and chemokines (Bilbo and Schwarz 2009), regulate the integrity of the BBB (Prat et al. 2001), and maintain brain homeostasis. Conversely, activated astrocytes and microglia can induce neuronal and synaptic changes, modifying CNS homeostasis, and contributing to neuronal dysfunction during disease. In postmortem brains of autistic patients, enhanced activation of astrocytes and microglia was observed in the subcortical white matter of the midfrontal gyrus, in the anterior cingulate gyrus, in the granular cell layer, in the Purkinje cell layer, and in the white matter of the cerebellum (Pardo et al. 2005; Vargas et al. 2005). Similar changes have been observed in animal models of autism (Lucchina and Depino 2014; more information in Chap. 8).

9.8 Neurotransmission Disturbed and Immune Dysfunction in ASD

Altered brain development can result in long-term changes in neuronal function, evidenced by altered neurotransmitter production and receptor expression. In addition, an alternative to the hypothesis of cytokines acting on the brain to induce behavioral changes is to propose that changes in brain development can result in persistent alterations of neurons, which can in turn modulate immune function. Differential changes in neurotransmitter systems have been referred in autism, including serotonin (5-HT), dopamine, noradrenaline (norepinephrine), gamma-aminobutryic acid (GABA), glutamate, and neuropeptides suggesting that a dysregulation in neurotransmission could result in altered development. Molecules like oxytocin and vasopressin are involved in immune tolerance, serotonin exerts suppressive or proliferative responses on T cells, and the vasoactive intestinal peptide (VIP) has an immunomodulatory action susceptible to be implicated in ASD.

It has been also reported that cytokines can affect enteric glial cells, although their effects on mood, sleep, nutritional intake, exploratory behavior, and social interactions do not necessarily imply that they cause neuronal damage or death (Bernardo et al. 2004). Systemic cytokine administration of IFN-α, IL-2, or TNF-α can cause increases in noradrenergic, dopaminergic, and serotonergic metabolism in the hypothalamus, hippocampus, and nuleus accumbens (Mohankumar et al. 1991; Shintani et al. 1993; Merali et al. 1997).

Neurotransmission at cholinergic synapses is essential for functional aspects of cognition, reward, motor activity and analgesia, and a dysregulation of *cholinergic neurotransmission* has been associated with autism. Differences in the regional expression of specific nicotinic acetylcholinergic receptor(nAChR) subunits in postmortem adult brains from autistic and non-autistic individuals have been reported (Ray et al. 2005). A decrease in the density of [3H]epibatidine binding sites within the parietal cortex, cerebellum, and thalamus of autistic brains was correlated to decreases in the levels of α3, α4, and β2 subunits (Martin-Ruiz et al. 2004; Lee et al. 2002). Although without a total understanding on the impact of this finding in autism pathology, acetylcholine receptor (AChR) expression raises important questions on whether deficits in nAChR neurotransmission identified in at least a subset of autistic children might confer a particularly high sensitivity to the adverse effects of organophosphates and/or neonicotinoid exposure. Indeed, allelic variants of the gene that encodes for a metabolic enzyme for the detoxification of organophosphates have been associated with autism in North America, and the ASD-related variant has lower metabolic activity in vitro (D'Amelio et al. 2005).

GABA receptors (GABAR) primarily mediate inhibitory responses, but also mediate excitatory transmission (Owens and Kriegstein 2002). This temporally regulated effect of GABA has been proposed to underlie a key role of this neurotransmitter in conveying information of the environment into a developing brain (Ben-Ari 2002). A failure in proper regulation of the chloride gradient can result in an imbalance between excitation and inhibition within developing neural circuits, a phenomenon that is proposed to underlie autistic symptoms (Snijders et al. 2013). Changes in GABA levels have been found in platelets of autistic children (Rolf et al. 1993). Moreover, the density of GABAR is diminished in the hippocampus of autistic individuals (Blatt et al. 2001), and both glutamic decarboxylases (GAD) 65 and GAD67 are reduced in the autistic parietal and cerebellar cortex (Fatemi et al. 2002). Differential levels of specific GABAA and GABAB receptors also suggest a role of GABA in autism (Schmitz et al. 2005; Fatemi et al. 2009a; Fatemi et al. 2009b).

Central catecholamine imbalance has been reported as a major factor in the behavioral phenotype in autism, and a hypofunction of dopamine (DA) and norepinephrine systems has been also suggested in autism. Attention to a role of DA in autism was drawn after the observation that some DA blockers (i.e., antipsychotics) can treat some aspects of autism (Arnold et al. 2010). However, the vast majority of studies measuring the levels of DA and its metabolites in the blood or the CSF of autistic individuals, have failed to detect differences with controls as was reviewed by Lam et al. (2006).

9.9 Current Directions on Immunomodulation in ASD

The concepts of autoimmunity and neuroimmune dysfunction in neurodevelopmental disorders are current important areas of research. However, some precautions are required in interpreting the information at this time of knowledge, since much of the evidences are not rigorous and most are aimed at identifying potential markers, with only preliminary evidences on the mechanisms underlying their involvement in the pathology of ASD.

Many of the experimental data mentioned in our review only suggest or support a primary role of failure of immune functions in the pathogenesis of this particular neuropsychiatric disorder, which does not imply that such a primary role could have been formally proved. We consider that some mechanisms underlying these links could be discovered using the recently developed and validated animal models (see Chap. 8).

This understanding may contribute to designing better protocols of treatment, useful to recover from neuronal dysfunction related to social interaction, verbal and nonverbal communication, as well as the repetitive, stereotypical, and overly restrictive behaviors present in ASD.

In the context of immunomodulatory therapy in ASD, it has been suggested that around 70 % of the children affected undergo complementary and alternative medicinal therapies to treat their symptoms (Levy et al. 2003; Wong and Smith 2006), including the use of probiotics, anti-infectives, and dietary augmentation (Golnik and Ireland 2009). The clinical trials aimed at testing therapies that modulate the immune response in children with ASD have addressed mainly the use of steroids, intravenous immunoglobulin G, antibiotics, and vitamin D (Cannell 2008; Gupta 1999; Plioplys 1998; Pradhan et al. 1999; Sandler et al. 2000; Dalakas 1997; Shenoy et al. 2000; Matarazzo 2002; Bradstreet et al. 2007; Buitelaar et al. 1990; Geier and Geier 2005). Understanding the biological effects of these treatments could probably provide a better background toward more effective conventional and alternative therapies.

9.10 Conclusions

Although it is well known that autism affects primarily the brain function (mainly social functioning and cognition), it is also known that it follows the compromise of other organs and systems. So, published evidences have identified widespread changes in the immune system of children with autism, both at systemic and cellular levels. Brain specimens from autistic subjects exhibit signs of active, ongoing inflammation, as well as alterations in gene pathways associated with immune signaling and immune function. Moreover, many genetic studies have also indicated a link between autism and relevant genes to both the nervous system and the immune system, which can in turn affect function in both systems. Together, these reports suggest that autism may in fact be a systemic disorder with connections to abnormal immune responses in and outside the brain. Consequently, the immune system dysfunction may represent a novel target for treatment. Hence, a better understanding of the involvement of the immune response in autism, and of how early brain development is altered, is crucial for a better and earlier intervention of the disease and has therapeutic implications. In our opinion, this chapter is a modest contribution to these propositions.

Acknowledgments This work was supported by Autism Biomarkers Project of CIREN (M.A.R-A.), Havana University (M.C.I.J.), a CONICET Grant PIP 2012, a University of Buenos Aires Grant UBACyT GEF2013-2015, and an ANPCyT Grant PICT2010 (A.M.D.). A.M.D. is a member of the Research Career at the National Council of Scientific and Technological Research (CONICET).

References

Abdallah MW, Larsen N, Grove J, Norgaard-Pedersen B, Thorsen P, Mortensen EL, Hougaard DM (2012) Amniotic fluid chemokines and autism spectrum disorders: an exploratory study utilizing a Danish Historic Birth Cohort. Brain Behav Immun 26(1):170–176. doi:S0889-1591(11)00521-6 [pii] 10.1016/j.bbi.2011.09.003

Abrahams VM, Schaefer TM, Fahey JV, Visintin I, Wright JA, Aldo PB, Romero R, Wira CR, Mor G (2006) Expression and secretion of antiviral factors by trophoblast cells following stimulation by the TLR-3 agonist, Poly(I: C). Hum Reprod 21(9):2432–2439. doi:10.1093/humrep/del178

Adams JB, Johansen LJ, Powell LD, Quig D, Rubin RA (2011) Gastrointestinal flora and gastrointestinal status in children with autism–comparisons to typical children and correlation with autism severity. BMC Gastroenterol 11:22. doi:10.1186/1471-230X-11-22

Afzal N, Murch S, Thirrupathy K, Berger L, Fagbemi A, Heuschkel R (2003) Constipation with acquired megarectum in children with autism. Pediatrics 112(4):939–942

Ahern PP, Izcue A, Maloy KJ, Powrie F (2008) The interleukin-23 axis in intestinal inflammation. Immunol Rev 226:147–159. doi:IMR705 [pii] 10.1111/j.1600-065X.2008.00705.x

Anderson GM, Freedman DX, Cohen DJ, Volkmar FR, Hoder EL, McPhedran P, Minderaa RB, Hansen CR, Young JG (1987) Whole blood serotonin in autistic and normal subjects. J Child Psychol Psychiatry 28(6):885–900

Arnold LE, Farmer C, Kraemer HC, Davies M, Witwer A, Chuang S, DiSilvestro R, McDougle CJ, McCracken J, Vitiello B, Aman MG, Scahill L, Posey DJ, Swiezy NB (2010) Moderators, mediators, and other predictors of risperidone response in children with autistic disorder and irritability. J Child Adolesc Psychopharmacol 20(2):83–93. doi:10.1089/cap.2009.0022

Ashwood P, Wakefield AJ (2006) Immune activation of peripheral blood and mucosal CD3+ lymphocyte cytokine profiles in children with autism and gastrointestinal symptoms. J Neuroimmunol 173(1–2):126–134. doi:S0165-5728(05)00539-4 [pii] 10.1016/j.jneuroim.2005.12.007

Ashwood P, Anthony A, Pellicer AA, Torrente F, Walker-Smith JA, Wakefield AJ (2003) Intestinal lymphocyte populations in children with regressive autism: evidence for extensive mucosal immunopathology. J Clin Immunol 23(6):504–517. doi:474304 [pii]

Ashwood P, Wills S, Van de Water J (2006) The immune response in autism: a new frontier for autism research. J Leukoc Biol 80(1):1–15. doi:jlb.1205707 [pii] 10.1189/jlb.1205707

Ashwood P, Enstrom A, Krakowiak P, Hertz-Picciotto I, Hansen RL, Croen LA, Ozonoff S, Pessah IN, Van de Water J (2008a) Decreased transforming growth factor beta1 in autism: a potential link between immune dysregulation and impairment in clinical behavioral outcomes. J Neuroimmunol 204(1–2):149–153. doi:S0165-5728(08)00293-2 [pii] 10.1016/j.jneuroim.2008.07.006

Ashwood P, Enstrom A, Krakowiak P, Hertz-Picciotto I, Hansen RL, Croen LA, Ozonoff S, Pessah IN, Van de Water J (2008b) Decreased transforming growth factor beta1 in autism: a potential link between immune dysregulation and impairment in clinical behavioral outcomes. J Neuroimmunol 204(1–2):149–153. doi:S0165-5728(08)00293-2 [pii] 10.1016/j.jneuroim.2008.07.006

Ashwood P, Krakowiak P, Hertz-Picciotto I, Hansen R, Pessah I, Van de Water J (2011a) Elevated plasma cytokines in autism spectrum disorders provide evidence of immune dysfunction and are associated with impaired behavioral outcome. Brain Behav Immun 25(1):40–45. doi:S0889-1591(10)00428-9 [pii] 10.1016/j.bbi.2010.08.003

Ashwood P, Krakowiak P, Hertz-Picciotto I, Hansen R, Pessah I, Van de Water J (2011b) Elevated plasma cytokines in autism spectrum disorders provide evidence of immune dysfunction and are associated with impaired behavioral outcome. Brain Behav Immun 25(1):40–45. doi:S0889-1591(10)00428-9 [pii] 10.1016/j.bbi.2010.08.003

Atladottir HO, Thorsen P, Ostergaard L, Schendel DE, Lemcke S, Abdallah M, Parner ET (2010) Maternal infection requiring hospitalization during pregnancy and autism spectrum disorders. J Autism Dev Disord 40(12):1423–1430. doi:10.1007/s10803-010-1006-y

Ayyoub M, Raffin C, Valmori D (2012) Generation of Th17 from human naive CD4+ T cells preferentially occurs from FOXP3+ Tregs upon costimulation via CD28 or CD5. Blood 119(20):4810–4812; author reply 4812-4813. doi:119/20/4810 [pii] 10.1182/blood-2012-02-409722

Banks WA (2012) Brain meets body: the blood-brain barrier as an endocrine interface. Endocrinology 153(9):4111–4119. doi:en.2012-1435 [pii] 10.1210/en.2012-1435

Barak Y, Kimhi R, Stein D, Gutman J, Weizman A (1999) Autistic subjects with comorbid epilepsy: a possible association with viral infections. Child Psychiatry Hum Dev 29(3):245–251

Ben-Ari Y (2002) Excitatory actions of gaba during development: the nature of the nurture. Nat Rev Neurosci 3(9):728–739

Bernardo A, Ball C, Nolasco L, Moake JF, Dong JF (2004) Effects of inflammatory cytokines on the release and cleavage of the endothelial cell-derived ultralarge von Willebrand factor multimers under flow. Blood 104(1):100–106. doi:10.1182/blood-2004-01-0107 2004-01-0107 [pii]

Bettelli E, Korn T, Kuchroo VK (2007) Th17: the third member of the effector T cell trilogy. Curr Opin Immunol 19(6):652–657. doi:S0952-7915(07)00133-1 [pii] 10.1016/j.coi.2007.07.020

Bilbo SD, Schwarz JM (2009) Early-life programming of later-life brain and behavior: a critical role for the immune system. Front Behav Neurosci 3:14. doi:10.3389/neuro.08.014.2009

Blatt GJ, Fitzgerald CM, Guptill JT, Booker AB, Kemper TL, Bauman ML (2001) Density and distribution of hippocampal neurotransmitter receptors in autism: an autoradiographic study. J Autism Dev Disord 31(6):537–543

Bradstreet JJ, Smith S, Granpeesheh D, El-Dahr JM, Rossignol D (2007) Spironolactone might be a desirable immunologic and hormonal intervention in autism spectrum disorders. Med Hypotheses 68(5):979–987. doi:S0306-9877(06)00753-5 [pii] 10.1016/j.mehy.2006.10.015

Buie T, Campbell DB, Fuchs GJ 3rd, Furuta GT, Levy J, Vandewater J, Whitaker AH, Atkins D, Bauman ML, Beaudet AL, Carr EG, Gershon MD, Hyman SL, Jirapinyo P, Jyonouchi H, Kooros K, Kushak R, Levitt P, Levy SE, Lewis JD, Murray KF, Natowicz MR, Sabra A, Wershil BK, Weston SC, Zeltzer L, Winter H (2010) Evaluation, diagnosis, and treatment of gastrointestinal disorders in individuals with ASDs: a consensus report. Pediatrics 125(Suppl 1):S1–18. doi:10.1542/peds.2009-1878C

Buitelaar JK, van Engeland H, van Ree JM, de Wied D (1990) Behavioral effects of Org 2766, a synthetic analog of the adrenocorticotrophic hormone (4–9), in 14 outpatient autistic children. J Autism Dev Disord 20(4):467–478
Burgess NK, Sweeten TL, McMahon WM, Fujinami RS (2006) Hyperserotoninemia and altered immunity in autism. J Autism Dev Disord 36(5):697–704. doi:10.1007/s10803-006-0100-7
Butler MG, Dasouki MJ, Zhou XP, Talebizadeh Z, Brown M, Takahashi TN, Miles JH, Wang CH, Stratton R, Pilarski R, Eng C (2005) Subset of individuals with autism spectrum disorders and extreme macrocephaly associated with germline PTEN tumour suppressor gene mutations. J Med Genet 42(4):318–321. doi:10.1136/jmg.2004.024646
Cannell JJ (2008) Autism and vitamin D. Med Hypotheses 70(4):750–759. doi:S0306-9877(07)00537-3 [pii] 10.1016/j.mehy.2007.08.016
Cappiello A, Malison RT, McDougle CJ, Vegso SJ, Charney DS, Heninger GR, Price LH (1996) Seasonal variation in neuroendocrine and mood responses to i.v. L-tryptophan in depressed patients and healthy subjects. Neuropsychopharmacology 15(5):475–483. doi:S0893-133X(96)00057-7 [pii] 10.1016/S0893-133X(96)00057-7
Chakrabarti S, Fombonne E (2005) Pervasive developmental disorders in preschool children: confirmation of high prevalence. Am J Psychiatry 162(6):1133–1141. doi:162/6/1133 [pii] 10.1176/appi.ajp.162.6.1133
Chess S (1971) Autism in children with congenital rubella. J Autism Child Schizophr 1(1):33–47
Chez MG, Dowling T, Patel PB, Khanna P, Kominsky M (2007) Elevation of tumor necrosis factor-alpha in cerebrospinal fluid of autistic children. Pediatr Neurol 36(6):361–365. doi:S0887-8994(07)00062-8 [pii] 10.1016/j.pediatrneurol.2007.01.012
Chugani DC, Muzik O, Behen M, Rothermel R, Janisse JJ, Lee J, Chugani HT (1999) Developmental changes in brain serotonin synthesis capacity in autistic and nonautistic children. Ann Neurol 45(3):287–295
Ciaranello AL, Ciaranello RD (1995) The neurobiology of infantile autism. Annu Rev Neurosci 18:101–128. doi:10.1146/annurev.ne.18.030195.000533
Coe CL, Lubach GR (2005) Prenatal origins of individual variation in behavior and immunity. Neurosci Biobehav Rev 29(1):39–49. doi:S0149-7634(04)00149-6 [pii] 10.1016/j.neubiorev.2004.11.003
Cohly HH, Panja A (2005) Immunological findings in autism. Int Rev Neurobiol 71:317–341
Comi AM, Zimmerman AW, Frye VH, Law PA, Peeden JN (1999) Familial clustering of autoimmune disorders and evaluation of medical risk factors in autism. J Child Neurol 14(6):388–394
Correll PH, Morrison AC, Lutz MA (2004) Receptor tyrosine kinases and the regulation of macrophage activation. J Leukoc Biol 75(5):731–737. doi:10.1189/jlb.0703347jlb.0703347 [pii]
Coury DL, Ashwood P, Fasano A, Fuchs G, Geraghty M, Kaul A, Mawe G, Patterson P, Jones NE (2012) Gastrointestinal conditions in children with autism spectrum disorder: developing a research agenda. Pediatrics 130(Suppl 2):S160–168. doi:10.1542/peds.2012-0900N
Coutinho AM, Sousa I, Martins M, Correia C, Morgadinho T, Bento C, Marques C, Ataide A, Miguel TS, Moore JH, Oliveira G, Vicente AM (2007) Evidence for epistasis between SLC6A4 and ITGB3 in autism etiology and in the determination of platelet serotonin levels. Hum Genet 121(2):243–256. doi:10.1007/s00439-006-0301-3
D'Amelio M, Ricci I, Sacco R, Liu X, D'Agruma L, Muscarella LA, Guarnieri V, Militerni R, Bravaccio C, Elia M, Schneider C, Melmed R, Trillo S, Pascucci T, Puglisi-Allegra S, Reichelt KL, Macciardi F, Holden JJ, Persico AM (2005) Paraoxonase gene variants are associated with autism in North America, but not in Italy: possible regional specificity in gene-environment interactions. Mol Psychiatry 10(11):1006–1016. doi:10.1038/sj.mp.4001714
D'Eufemia P, Celli M, Finocchiaro R, Pacifico L, Viozzi L, Zaccagnini M, Cardi E, Giardini O (1996) Abnormal intestinal permeability in children with autism. Acta Paediatr 85(9):1076–1079
Dalakas MC (1997) Intravenous immune globulin therapy for neurologic diseases. Ann Intern Med 126(9):721–730
Daniels WW, Warren RP, Odell JD, Maciulis A, Burger RA, Warren WL, Torres AR (1995) Increased frequency of the extended or ancestral haplotype B44-SC30-DR4 in autism. Neuropsychobiology 32(3):120–123

de Magistris L, Familiari V, Pascotto A, Sapone A, Frolli A, Iardino P, Carteni M, De Rosa M, Francavilla R, Riegler G, Militerni R, Bravaccio C (2010) Alterations of the intestinal barrier in patients with autism spectrum disorders and in their first-degree relatives. J Pediatr Gastroenterol Nutr 51(4):418–424. doi:10.1097/MPG.0b013e3181dcc4a5

de Theije CG, Wu J, da Silva SL, Kamphuis PJ, Garssen J, Korte SM, Kraneveld AD (2011) Pathways underlying the gut-to-brain connection in autism spectrum disorders as future targets for disease management. Eur J Pharmacol 668(Suppl 1):S70–80. doi:10.1016/j.ejphar.2011.07.013

DeLong GR, Bean SC, Brown FR 3rd (1981) Acquired reversible autistic syndrome in acute encephalopathic illness in children. Arch Neurol 38(3):191–194

Depino AM (2006) Maternal infection and the offspring brain. J Neurosci 26(30):7777–7778

Depino AM (2013) Peripheral and central inflammation in autism spectrum disorders. Mol Cell Neurosci 53:69–76. doi:S1044-7431(12)00185-6 [pii] 10.1016/j.mcn.2012.10.003

Depino AM, Alonso M, Ferrari C, del Rey A, Anthony D, Besedovsky H, Medina JH, Pitossi F (2004) Learning modulation by endogenous hippocampal IL-1: blockade of endogenous IL-1 facilitates memory formation. Hippocampus 14(4):526–535

Depino AM, Lucchina L, Pitossi F (2011) Early and adult hippocampal TGF-beta1 overexpression have opposite effects on behavior. Brain Behav Immun 25(8):1582–1591. doi:S0889-1591(11)00180-2 [pii] 10.1016/j.bbi.2011.05.007

Deverman BE, Patterson PH (2009) Cytokines and CNS development. Neuron 64(1):61–78. doi:S0896-6273(09)00680-1 [pii] 10.1016/j.neuron.2009.09.002

Dong C (2008) TH17 cells in development: an updated view of their molecular identity and genetic programming. Nat Rev Immunol 8(5):337–348. doi:nri2295 [pii] 10.1038/nri2295

Emanuele E, Orsi P, Boso M, Broglia D, Brondino N, Barale F, di Nemi SU, Politi P (2010) Low-grade endotoxemia in patients with severe autism. Neurosci Lett 471(3):162–165. doi:S0304-3940(10)00061-3 [pii] 10.1016/j.neulet.2010.01.033

Engstrom H, Ohlson S, Stubbs EG, Maciulis A, Caldwell V, Odell JD, Torres A (2003) Decreased expression of CD95 (FAS/APO-1) on CD4+ T-lymphocytes from participants with autism. J Develop Phys Disabil 15(2):155–163. doi:10.1023/A:1022827417414

Enstrom AM, Onore CE, de Water JAV, Ashwood P (2010a) Differential monocyte responses to TLR ligands in children with autism spectrum disorders. Brain Behav Immun 24(1):64–71. doi:10.1016/j.bbi.2009.08.001

Enstrom AM, Onore CE, Van de Water JA, Ashwood P (2010b) Differential monocyte responses to TLR ligands in children with autism spectrum disorders. Brain Behav Immun 24(1):64–71. doi:S0889-1591(09)00388-2 [pii] 10.1016/j.bbi.2009.08.001

Erickson MA, Dohi K, Banks WA (2012) Neuroinflammation: a common pathway in CNS diseases as mediated at the blood-brain barrier. Neuroimmunomodulation 19(2):121–130. doi:000330247 [pii] 10.1159/000330247

Fatemi SH, Stary JM, Halt AR, Realmuto GR (2001) Dysregulation of Reelin and Bcl-2 proteins in autistic cerebellum. J Autism Dev Disord 31(6):529–535

Fatemi SH, Halt AR, Stary JM, Kanodia R, Schulz SC, Realmuto GR (2002) Glutamic acid decarboxylase 65 and 67 kDa proteins are reduced in autistic parietal and cerebellar cortices. Biol Psychiatry 52(8):805–810

Fatemi SH, Folsom TD, Reutiman TJ, Thuras PD (2009a) Expression of GABA(B) receptors is altered in brains of subjects with autism. Cerebellum 8(1):64–69. doi:10.1007/s12311-008-0075-3

Fatemi SH, Reutiman TJ, Folsom TD, Thuras PD (2009b) GABA(A) receptor downregulation in brains of subjects with autism. J Autism Dev Disord 39(2):223–230. doi:10.1007/s10803-008-0646-7

Finegold SM (2011) State of the art; microbiology in health and disease. Intestinal bacterial flora in autism. Anaerobe 17(6):367–368. doi:10.1016/j.anaerobe.2011.03.007

Finegold SM, Molitoris D, Song Y, Liu C, Vaisanen ML, Bolte E, McTeague M, Sandler R, Wexler H, Marlowe EM, Collins MD, Lawson PA, Summanen P, Baysallar M, Tomzynski TJ, Read E, Johnson E, Rolfe R, Nasir P, Shah H, Haake DA, Manning P, Kaul A (2002) Gastrointestinal microflora studies in late-onset autism. Clin Infect Dis 35(Suppl 1):S6–S16. doi:10.1086/341914

Finegold SM, Dowd SE, Gontcharova V, Liu C, Henley KE, Wolcott RD, Youn E, Summanen PH, Granpeesheh D, Dixon D, Liu M, Molitoris DR, Green JA 3rd (2010) Pyrosequencing study of fecal microflora of autistic and control children. Anaerobe 16(4):444–453. doi:10.1016/j.anaerobe.2010.06.008

Finegold SM, Downes J, Summanen PH (2012) Microbiology of regressive autism. Anaerobe 18(2):260–262. doi:10.1016/j.anaerobe.2011.12.018

Fombonne E (2005) Epidemiology of autistic disorder and other pervasive developmental disorders. J Clin Psychiatry 66(Suppl 10):3–8

Furlano RI, Anthony A, Day R, Brown A, McGarvey L, Thomson MA, Davies SE, Berelowitz M, Forbes A, Wakefield AJ, Walker-Smith JA, Murch SH (2001) Colonic CD8 and gamma delta T-cell infiltration with epithelial damage in children with autism. J Pediatr 138(3):366–372. doi:S0022-3476(01)92227-9 [pii] 10.1067/mpd.2001.111323

Garay PA, Hsiao EY, Patterson PH, McAllister AK (2012) Maternal immune activation causes age- and region-specific changes in brain cytokines in offspring throughout development. Brain Behav Immun. doi:S0889-1591(12)00191-2 [pii] 10.1016/j.bbi.2012.07.008

Garbett K, Ebert PJ, Mitchell A, Lintas C, Manzi B, Mirnics K, Persico AM (2008) Immune transcriptome alterations in the temporal cortex of subjects with autism. Neurobiol Dis 30(3):303–311. doi:S0969-9961(08)00024-7 [pii] 10.1016/j.nbd.2008.01.012

Geier MR, Geier DA (2005) The potential importance of steroids in the treatment of autistic spectrum disorders and other disorders involving mercury toxicity. Med Hypotheses 64(5):946–954. doi:S0306-9877(04)00615-2 [pii] 10.1016/j.mehy.2004.11.018

Ghaziuddin M, Tsai LY, Eilers L, Ghaziuddin N (1992) Brief report: autism and herpes simplex encephalitis. J Autism Dev Disord 22(1):107–113

Gillberg C (1986) Onset at age 14 of a typical autistic syndrome. A case report of a girl with herpes simplex encephalitis. J Autism Dev Disord 16(3):369–375

Golnik AE, Ireland M (2009) Complementary alternative medicine for children with autism: a physician survey. J Autism Dev Disord 39(7):996–1005. doi:10.1007/s10803-009-0714-7

Greer MK, Lyons-Crews M, Mauldin LB, Brown FR 3rd (1989) A case study of the cognitive and behavioral deficits of temporal lobe damage in herpes simplex encephalitis. J Autism Dev Disord 19(2):317–326

Gregg JP, Lit L, Baron CA, Hertz-Picciotto I, Walker W, Davis RA, Croen LA, Ozonoff S, Hansen R, Pessah IN, Sharp FR (2008) Gene expression changes in children with autism. Genomics 91(1):22–29. doi:S0888-7543(07)00232-7 [pii] 10.1016/j.ygeno.2007.09.003

Grigorenko EL, Han SS, Yrigollen CM, Leng L, Mizue Y, Anderson GM, Mulder EJ, de Bildt A, Minderaa RB, Volkmar FR, Chang JT, Bucala R (2008) Macrophage migration inhibitory factor and autism spectrum disorders. Pediatrics 122(2):e438–445. doi:122/2/e438 [pii] 10.1542/peds.2007-3604

Gupta S (1999) Treatment of children with autism with intravenous immunoglobulin. J Child Neurol 14(3):203–205

Gupta S, Aggarwal S, Rashanravan B, Lee T (1998) Th1- and Th2-like cytokines in CD4+ and CD8+ T cells in autism. J Neuroimmunol 85(1):106–109. doi:S0165-5728(98)00021-6 [pii]

Hanley HG, Stahl SM, Freedman DX (1977) Hyperserotonemia and amine metabolites in autistic and retarded children. Arch Gen Psychiatry 34(5):521–531

Harrington LE, Mangan PR, Weaver CT (2006) Expanding the effector CD4 T-cell repertoire: the Th17 lineage. Curr Opin Immunol 18(3):349–356. doi:S0952-7915(06)00065-3 [pii] 10.1016/j.coi.2006.03.017

Herman GE, Butter E, Enrile B, Pastore M, Prior TW, Sommer A (2007) Increasing knowledge of PTEN germline mutations: two additional patients with autism and macrocephaly. Am J Med Genet A 143(6):589–593. doi:10.1002/ajmg.a.31619

Hertz-Picciotto I, Croen LA, Hansen R, Jones CR, van de Water J, Pessah IN (2006) The CHARGE study: an epidemiologic investigation of genetic and environmental factors contributing to autism. Environ Health Perspect 114(7):1119–1125

Hoeffer CA, Klann E (2010) mTOR signaling: at the crossroads of plasticity, memory and disease. Trends Neurosci 33(2):67–75. doi:10.1016/j.tins.2009.11.003

Hoekstra RA, Wheelwright S (2009) Autistic traits in simplex and multiplex autism families: focus on unaffected relatives. Am J Med Genet B Neuropsychiatr Genet. doi:10.1002/ajmg.b.30947

Hoekstra PJ, Troost PW, Lahuis BE, Mulder H, Mulder EJ, Franke B, Buitelaar JK, Anderson GM, Scahill L, Minderaa RB (2010) Risperidone-induced weight gain in referred children with autism spectrum disorders is associated with a common polymorphism in the 5-hydroxytryptamine 2 C receptor gene. J Child Adolesc Psychopharmacol 20(6):473–477. doi:10.1089/cap.2009.0071

Holt R, Barnby G, Maestrini E, Bacchelli E, Brocklebank D, Sousa I, Mulder EJ, Kantojarvi K, Jarvela I, Klauck SM, Poustka F, Bailey AJ, Monaco AP (2010) Linkage and candidate gene studies of autism spectrum disorders in European populations. Eur J Hum Genet 18(9):1013–1019. doi:ejhg201069 [pii] 10.1038/ejhg.2010.69

Horvath K, Perman JA (2002) Autism and gastrointestinal symptoms. Curr Gastroenterol Rep 4(3):251–258

Hoyler T, Klose CS, Souabni A, Turqueti-Neves A, Pfeifer D, Rawlins EL, Voehringer D, Busslinger M, Diefenbach A (2012) The transcription factor GATA-3 controls cell fate and maintenance of type 2 innate lymphoid cells. Immunity 37(4):634–648. doi:10.1016/j.immuni.2012.06.020

Ivarsson SA, Bjerre I, Vegfors P, Ahlfors K (1990) Autism as one of several disabilities in two children with congenital cytomegalovirus infection. Neuropediatrics 21(2):102–103. doi:10.1055/s-2008-1071471

Jyonouchi H, Geng L, Ruby A, Zimmerman-Bier B (2005) Dysregulated innate immune responses in young children with autism spectrum disorders: their relationship to gastrointestinal symptoms and dietary intervention. Neuropsychobiology 51(2):77–85. doi:84164 [pii] 10.1159/000084164

Jyonouchi H, Geng L, Cushing-Ruby A, Quraishi H (2008) Impact of innate immunity in a subset of children with autism spectrum disorders: a case control study. J Neuroinflammation 5:52. doi:1742-2094-5-52 [pii] 10.1186/1742-2094-5-52

Kang DW, Park JG, Ilhan ZE, Wallstrom G, Labaer J, Adams JB, Krajmalnik-Brown R (2013) Reduced incidence of Prevotella and other fermenters in intestinal microflora of autistic children. PLoS ONE 8(7):e68322. doi:10.1371/journal.pone.0068322

Kersse K, Vanden Berghe T, Lamkanfi M, Vandenabeele P (2007) A phylogenetic and functional overview of inflammatory caspases and caspase-1-related CARD-only proteins. Biochem Soc Trans 35(Pt 6):1508–1511. doi:BST0351508 [pii] 10.1042/BST0351508

Konsman JP, Parnet P, Dantzer R (2002) Cytokine-induced sickness behaviour: mechanisms and implications. Trends Neurosci 25(3):154–159. doi:S0166223600020889 [pii]

Lam KS, Aman MG, Arnold LE (2006) Neurochemical correlates of autistic disorder: a review of the literature. Res Dev Disabil 27(3):254–289. doi:10.1016/j.ridd.2005.03.003

Lamkanfi M, Kanneganti TD, Franchi L, Nunez G (2007) Caspase-1 inflammasomes in infection and inflammation. J Leukoc Biol 82(2):220–225. doi:jlb.1206756 [pii] 10.1189/jlb.1206756

Lee M, Martin-Ruiz C, Graham A, Court J, Jaros E, Perry R, Iversen P, Bauman M, Perry E (2002) Nicotinic receptor abnormalities in the cerebellar cortex in autism. Brain 125(Pt 7):1483–1495

Lee LC, Zachary AA, Leffell MS, Newschaffer CJ, Matteson KJ, Tyler JD, Zimmerman AW (2006) HLA-DR4 in families with autism. Pediatr Neurol 35(5):303–307. doi:S0887-8994(06)00383-3 [pii] 10.1016/j.pediatrneurol.2006.06.006

Letterio JJ, Roberts AB (1998) Regulation of immune responses by TGF-beta. Annu Rev Immunol 16:137–161. doi:10.1146/annurev.immunol.16.1.137

Levy SE, Mandell DS, Merhar S, Ittenbach RF, Pinto-Martin JA (2003) Use of complementary and alternative medicine among children recently diagnosed with autistic spectrum disorder. J Dev Behav Pediatr 24(6):418–423

Li X, Chauhan A, Sheikh AM, Patil S, Chauhan V, Li XM, Ji L, Brown T, Malik M (2009) Elevated immune response in the brain of autistic patients. J Neuroimmunol 207(1–2):111–116. doi:S0165-5728(08)00490-6 [pii] 10.1016/j.jneuroim.2008.12.002

Lintas C, Sacco R, Garbett K, Mirnics K, Militerni R, Bravaccio C, Curatolo P, Manzi B, Schneider C, Melmed R, Elia M, Pascucci T, Puglisi-Allegra S, Reichelt KL, Persico AM (2009) Involvement of the PRKCB1 gene in autistic disorder: significant genetic association and reduced neocortical gene expression. Mol Psychiatry 14(7):705–718. doi:mp200821 [pii] 10.1038/mp.2008.21

Liu G, Yang K, Burns S, Shrestha S, Chi H (2010) The S1P(1)-mTOR axis directs the reciprocal differentiation of T(H)1 and T(reg) cells. Nat Immunol 11(11):1047–1056. doi:ni.1939 [pii] 10.1038/ni.1939

Lucchina L, Depino AM (2014) Altered peripheral and central inflammatory responses in a mouse model of autism. Autism Res 7(2):273–289. doi:10.1002/aur.1338

MacFabe DF, Cain NE, Boon F, Ossenkopp KP, Cain DP (2011) Effects of the enteric bacterial metabolic product propionic acid on object-directed behavior, social behavior, cognition, and neuroinflammation in adolescent rats: Relevance to autism spectrum disorder. Behav Brain Res 217(1):47–54. doi:S0166-4328(10)00677-7 [pii] 10.1016/j.bbr.2010.10.005

Madsen-Bouterse SA, Romero R, Tarca AL, Kusanovic JP, Espinoza J, Kim CJ, Kim JS, Edwin SS, Gomez R, Draghici S (2010) The transcriptome of the fetal inflammatory response syndrome. Am J Reprod Immunol 63(1):73–92. doi:AJI791 [pii] 10.1111/j.1600-0897.2009.00791.x

Malik M, Sheikh AM, Wen G, Spivack W, Brown WT, Li X (2011) Expression of inflammatory cytokines, Bcl2 and cathepsin D are altered in lymphoblasts of autistic subjects. Immunobiology 216(1–2):80–85. doi:S0171-2985(10)00024-0 [pii] 10.1016/j.imbio.2010.03.001

Maloy KJ (2008) The Interleukin-23/Interleukin-17 axis in intestinal inflammation. J Intern Med 263(6):584–590. doi:10.1111/j.1365-2796.2008.01950.x

Markowitz PI (1983) Autism in a child with congenital cytomegalovirus infection. J Autism Dev Disord 13(3):249–253

Martin-Ruiz CM, Lee M, Perry RH, Baumann M, Court JA, Perry EK (2004) Molecular analysis of nicotinic receptor expression in autism. Brain Res Mol Brain Res 123(1–2):81–90. doi:10.1016/j.molbrainres.2004.01.003S0169328X0400035X [pii]

Matarazzo EB (2002) Treatment of late onset autism as a consequence of probable autommune processes related to chronic bacterial infection. World J Biol Psychiatry 3(3):162–166

McDougle CJ, Naylor ST, Cohen DJ, Aghajanian GK, Heninger GR, Price LH (1996) Effects of tryptophan depletion in drug-free adults with autistic disorder. Arch Gen Psychiatry 53(11):993–1000

McLay RN, Kastin AJ, Zadina JE (2000) Passage of interleukin-1-beta across the blood-brain barrier is reduced in aged mice: a possible mechanism for diminished fever in aging. Neuroimmunomodulation 8(3):148–153. doi:54275 [pii] 54275

Medzhitov R, Janeway C Jr (2000) Innate immunity. N Engl J Med 343(5):338–344. doi:10.1056/NEJM200008033430506

Merali Z, Lacosta S, Anisman H (1997) Effects of interleukin-1beta and mild stress on alterations of norepinephrine, dopamine and serotonin neurotransmission: a regional microdialysis study. Brain Res 761(2):225–235. doi:S0006-8993(97)00312-0 [pii]

Merrill JE (1992) Tumor necrosis factor alpha, interleukin 1 and related cytokines in brain development: normal and pathological. Dev Neurosci 14(1):1–10

Meyer U, Nyffeler M, Engler A, Urwyler A, Schedlowski M, Knuesel I, Yee BK, Feldon J (2006) The time of prenatal immune challenge determines the specificity of inflammation-mediated brain and behavioral pathology. J Neurosci 26(18):4752–4762

Meyer U, Yee BK, Feldon J (2007) The neurodevelopmental impact of prenatal infections at different times of pregnancy: the earlier the worse? Neuroscientist 13(3):241–256. doi:13/3/241 [pii] 10.1177/1073858406296401

Meyer U, Feldon J, Yee BK (2009) A review of the fetal brain cytokine imbalance hypothesis of schizophrenia. Schizophr Bull 35(5):959–972. doi:sbn022 [pii] 10.1093/schbul/sbn022

Minderaa RB, Anderson GM, Volkmar FR, Akkerhuis GW, Cohen DJ (1987) Urinary 5-hydroxyindoleacetic acid and whole blood serotonin and tryptophan in autistic and normal subjects. Biol Psychiatry 22(8):933–940

Mjosberg J, Bernink J, Golebski K, Karrich JJ, Peters CP, Blom B, te Velde AA, Fokkens WJ, van Drunen CM, Spits H (2012) The transcription factor GATA3 is essential for the function of human type 2 innate lymphoid cells. Immunity 37(4):649–659. doi:10.1016/j.immuni.2012.08.015

Mohankumar PS, Thyagarajan S, Quadri SK (1991) Interleukin-1 stimulates the release of dopamine and dihydroxyphenylacetic acid from the hypothalamus in vivo. Life Sci 48(9):925–930

Molloy CA, Morrow AL, Meinzen-Derr J, Schleifer K, Dienger K, Manning-Courtney P, Altaye M, Wills-Karp M (2006) Elevated cytokine levels in children with autism spectrum disorder. J Neuroimmunol 172(1–2):198–205. doi:S0165-5728(05)00492-3 [pii] 10.1016/j.jneuroim.2005.11.007

Mondino A, Mueller DL (2007) mTOR at the crossroads of T cell proliferation and tolerance. Semin Immunol 19(3):162–172. doi:10.1016/j.smim.2007.02.008

Mostafa GA, Al Shehab A, Fouad NR (2010) Frequency of CD4+ CD25high regulatory T cells in the peripheral blood of Egyptian children with autism. J Child Neurol 25(3):328–335. doi:10.1177/0883073809339393

Niehus R, Lord C (2006) Early medical history of children with autism spectrum disorders. J Dev Behav Pediatr 27(2 Suppl):S120–127

Nijmeijer JS, Arias-Vasquez A, Rommelse NN, Altink ME, Anney RJ, Asherson P, Banaschewski T, Buschgens CJ, Fliers EA, Gill M, Minderaa RB, Poustka L, Sergeant JA, Buitelaar JK, Franke B, Ebstein RP, Miranda A, Mulas F, Oades RD, Roeyers H, Rothenberger A, Sonuga-Barke EJ, Steinhausen HC, Faraone SV, Hartman CA, Hoekstra PJ (2010) Identifying loci for the overlap between attention-deficit/hyperactivity disorder and autism spectrum disorder using a genome-wide QTL linkage approach. J Am Acad Child Adolesc Psychiatry 49(7):675–685. doi:10.1016/j.jaac.2010.03.015

Okada K, Hashimoto K, Iwata Y, Nakamura K, Tsujii M, Tsuchiya KJ, Sekine Y, Suda S, Suzuki K, Sugihara G, Matsuzaki H, Sugiyama T, Kawai M, Minabe Y, Takei N, Mori N (2007) Decreased serum levels of transforming growth factor-beta1 in patients with autism. Prog Neuropsychopharmacol Biol Psychiatry 31(1):187–190. doi:S0278-5846(06)00339-3 [pii] 10.1016/j.pnpbp.2006.08.020

Olszak T, An D, Zeissig S, Vera MP, Richter J, Franke A, Glickman JN, Siebert R, Baron RM, Kasper DL, Blumberg RS (2012) Microbial exposure during early life has persistent effects on natural killer T cell function. Science 336(6080):489–493. doi:10.1126/science.1219328

Onore C, Enstrom A, Krakowiak P, Hertz-Picciotto I, Hansen R, Van de Water J, Ashwood P (2009) Decreased cellular IL-23 but not IL-17 production in children with autism spectrum disorders. J Neuroimmunol 216(1–2):126–129. doi:10.1016/j.jneuroim.2009.09.005

Owens DF, Kriegstein AR (2002) Is there more to GABA than synaptic inhibition? Nat Rev Neurosci 3(9):715–727

Pardo CA, Eberhart CG (2007) The neurobiology of autism. Brain Pathol 17(4):434–447

Pardo CA, Vargas DL, Zimmerman AW (2005) Immunity, neuroglia and neuroinflammation in autism. Int Rev Psychiatry (Abingdon, England) 17(6):485–495

Parracho HM, Bingham MO, Gibson GR, McCartney AL (2005) Differences between the gut microflora of children with autistic spectrum disorders and that of healthy children. J Med Microbiol 54(Pt 10):987–991. doi:54/10/987 [pii] 10.1099/jmm.0.46101-0

Plioplys AV (1998) Intravenous immunoglobulin treatment of children with autism. J Child Neurol 13(2):79–82

Pradhan S, Gupta RP, Shashank S, Pandey N (1999) Intravenous immunoglobulin therapy in acute disseminated encephalomyelitis. J Neurol Sci 165(1):56–61. doi:S0022-510X(99)00072-6 [pii]

Prat A, Biernacki K, Wosik K, Antel JP (2001) Glial cell influence on the human blood-brain barrier. Glia 36(2):145–155. doi:10.1002/glia.1104 [pii]

Ray MA, Graham AJ, Lee M, Perry RH, Court JA, Perry EK (2005) Neuronal nicotinic acetylcholine receptor subunits in autism: an immunohistochemical investigation in the thalamus. Neurobiol Dis 19(3):366–377. doi:10.1016/j.nbd.2005.01.017

Ring A, Barak Y, Ticher A, Ashkenazi I, Elizur A, Weizman A (1997) Evidence for an infectious etiology in autism. Pathophysiology 4:91–96

Rodriguez-Escudero I, Oliver MD, Andres-Pons A, Molina M, Cid VJ, Pulido R (2011) A comprehensive functional analysis of PTEN mutations: implications in tumor- and autism-related syndromes. Hum Mol Genet 20(21):4132–4142. doi:10.1093/hmg/ddr337

Rolf LH, Haarmann FY, Grotemeyer KH, Kehrer H (1993) Serotonin and amino acid content in platelets of autistic children. Acta Psychiatr Scand 87(5):312–316

Romagnani S (2000) T-cell subsets (Th1 versus Th2). Ann Allergy Asthma Immunol 85(1):9–18; quiz 18, 21. doi:S1081-1206(10)62426-X [pii] 10.1016/S1081-1206(10)62426-X

Romagnani P, Annunziato F, Piccinni MP, Maggi E, Romagnani S (2000) Th1/Th2 cells, their associated molecules and role in pathophysiology. Eur Cytokine Netw 11(3):510–511

Romero R, Gotsch F, Pineles B, Kusanovic JP (2007) Inflammation in pregnancy: its roles in reproductive physiology, obstetrical complications, and fetal injury. Nutr Rev 65(12 Pt 2):S194–202

Rosen NJ, Yoshida CK, Croen LA (2007) Infection in the first 2 years of life and autism spectrum disorders. Pediatrics 119(1):e61–69. doi:119/1/e61 [pii] 10.1542/peds.2006-1788

Round JL, O'Connell RM, Mazmanian SK (2010) Coordination of tolerogenic immune responses by the commensal microbiota. J Autoimmun 34(3):J220–225. doi:10.1016/j.jaut.2009.11.007

Rout UK, Clausen P (2009) Common increase of GATA-3 level in PC-12 cells by three teratogens causing autism spectrum disorders. Neurosci Res 64(2):162–169. doi:S0168-0102(09)00063-7 [pii] 10.1016/j.neures.2009.02.009

Sandler RH, Finegold SM, Bolte ER, Buchanan CP, Maxwell AP, Vaisanen ML, Nelson MN, Wexler HM (2000) Short-term benefit from oral vancomycin treatment of regressive-onset autism. J Child Neurol 15(7):429–435

Sauer S, Bruno L, Hertweck A, Finlay D, Leleu M, Spivakov M, Knight ZA, Cobb BS, Cantrell D, O'Connor E, Shokat KM, Fisher AG, Merkenschlager M (2008) T cell receptor signaling controls Foxp3 expression via PI3K, Akt, and mTOR. Proc Natl Acad Sci U S A 105(22):7797–7802. doi:10.1073/pnas.0800928105

Schain RJ, Freedman DX (1961) Studies on 5-hydroxyindole metabolism in autistic and other mentally retarded children. J Pediatr 58:315–320

Schmitz C, van Kooten IA, Hof PR, van Engeland H, Patterson PH, Steinbusch HW (2005) Autism: neuropathology, alterations of the GABAergic system, and animal models. Int Rev Neurobiol 71:1–26

Shenoy S, Arnold S, Chatila T (2000) Response to steroid therapy in autism secondary to autoimmune lymphoproliferative syndrome. J Pediatr 136(5):682–687. doi:S0022-3476(00)48959-6 [pii] 10.1067/mpd.2000.105355

Shintani F, Kanba S, Nakaki T, Nibuya M, Kinoshita N, Suzuki E, Yagi G, Kato R, Asai M (1993) Interleukin-1 beta augments release of norepinephrine, dopamine, and serotonin in the rat anterior hypothalamus. J Neurosci 13(8):3574–3581

Singh VK (1996) Plasma increase of interleukin-12 and interferon-gamma. Pathological significance in autism. J Neuroimmunol 66(1–2):143–145

Singh VK, Jensen RL (2003) Elevated levels of measles antibodies in children with autism. Pediatr Neurol 28(4):292–294. doi:S0887899402006276 [pii]

Siniscalco D, Antonucci N (2013) Involvement of dietary bioactive proteins and peptides in autism spectrum disorders. Curr Protein Pept Sci 14(8):674–679

Siniscalco D, Sapone A, Giordano C, Cirillo A, de Novellis V, de Magistris L, Rossi F, Fasano A, Maione S, Antonucci N (2012) The expression of caspases is enhanced in peripheral blood mononuclear cells of autism spectrum disorder patients. J Autism Dev Disord 42(7):1403–1410. doi:10.1007/s10803-011-1373-z

Snijders TM, Milivojevic B, Kemner C (2013) Atypical excitation-inhibition balance in autism captured by the gamma response to contextual modulation. Neuroimage Clin 3:65–72. doi:10.1016/j.nicl.2013.06.015

Sousa I, Clark TG, Holt R, Pagnamenta AT, Mulder EJ, Minderaa RB, Bailey AJ, Battaglia A, Klauck SM, Poustka F, Monaco AP (2010) Polymorphisms in leucine-rich repeat genes are associated with autism spectrum disorder susceptibility in populations of European ancestry. Mol Autism 1(1):7. doi:2040-2392-1-7 [pii] 10.1186/2040-2392-1-7

Stubbs EG (1978) Autistic symptoms in a child with congenital cytomegalovirus infection. J Autism Child Schizophr 8(1):37–43

Stubbs EG, Ash E, Williams CP (1984) Autism and congenital cytomegalovirus. J Autism Dev Disord 14(2):183–189

Sun CM, Hall JA, Blank RB, Bouladoux N, Oukka M, Mora JR, Belkaid Y (2007) Small intestine lamina propria dendritic cells promote de novo generation of Foxp3 T reg cells via retinoic acid. J Exp Med 204(8):1775–1785. doi:jem.20070602 [pii] 10.1084/jem.20070602

Suzuki K, Matsuzaki H, Iwata K, Kameno Y, Shimmura C, Kawai S, Yoshihara Y, Wakuda T, Takebayashi K, Takagai S, Matsumoto K, Tsuchiya KJ, Iwata Y, Nakamura K, Tsujii M, Sugiyama T, Mori N (2011) Plasma cytokine profiles in subjects with high-functioning autism spectrum disorders. PLoS One 6(5):e20470. doi:10.1371/journal.pone.0020470 PONE-D-11-06135 [pii]

Torrente F, Ashwood P, Day R, Machado N, Furlano RI, Anthony A, Davies SE, Wakefield AJ, Thomson MA, Walker-Smith JA, Murch SH (2002) Small intestinal enteropathy with epithelial IgG and complement deposition in children with regressive autism. Mol Psychiatry 7(4):375–382, 334. doi:10.1038/sj.mp.4001077

Torres AR, Sweeten TL, Cutler A, Bedke BJ, Fillmore M, Stubbs EG, Odell D (2006) The association and linkage of the HLA-A2 class I allele with autism. Hum Immunol 67(4–5):346–351. doi:S0198-8859(06)00042-5 [pii] 10.1016/j.humimm.2006.01.001

Ueda A, Zhou L, Stein PL (2012) Fyn promotes Th17 differentiation by regulating the kinetics of RORgammat and Foxp3 expression. J Immunol 188(11):5247–5256. doi:jimmunol.1102241 [pii] 10.4049/jimmunol.1102241

van Doorninck JH, van Der Wees J, Karis A, Goedknegt E, Engel JD, Coesmans M, Rutteman M, Grosveld F, De Zeeuw CI (1999) GATA-3 is involved in the development of serotonergic neurons in the caudal raphe nuclei. J Neurosci 19(12):RC12

van Gent T, Heijnen CJ, Treffers PD (1997) Autism and the immune system. J Child Psychol Psychiatry 38(3):337–349

Vargas DL, Nascimbene C, Krishnan C, Zimmerman AW, Pardo CA (2005) Neuroglial activation and neuroinflammation in the brain of patients with autism. Ann Neurol 57(1):67–81

Vojdani A, Pangborn JB, Vojdani E, Cooper EL (2003) Infections, toxic chemicals and dietary peptides binding to lymphocyte receptors and tissue enzymes are major instigators of autoimmunity in autism. Int J Immunopathol Pharmacol 16(3):189–199

Wakefield AJ, Anthony A, Murch SH, Thomson M, Montgomery SM, Davies S, O'Leary JJ, Berelowitz M, Walker-Smith JA (2000) Enterocolitis in children with developmental disorders. Am J Gastroenterol 95(9):2285–2295. doi:S0002-9270(00)02041-4 [pii] 10.1111/j.1572-0241.2000.03248.x

Wallach D, Kang TB, Rajput A, Kim JC, Bogdanov K, Yang SH, Kovalenko A (2010) Anti-inflammatory functions of the "apoptotic" caspases. Ann N Y Acad Sci 1209:17–22. doi:10.1111/j.1749-6632.2010.05742.x

Warren RP, Margaretten NC, Pace NC, Foster A (1986) Immune abnormalities in patients with autism. J Autism Dev Disord 16(2):189–197

Warren RP, Yonk J, Burger RW, Odell D, Warren WL (1995) DR-positive T cells in autism: association with decreased plasma levels of the complement C4B protein. Neuropsychobiology 31(2):53–57

Wong HH, Smith RG (2006) Patterns of complementary and alternative medical therapy use in children diagnosed with autism spectrum disorders. J Autism Dev Disord 36(7):901–909. doi:10.1007/s10803-006-0131-0

Yamashita Y, Fujimoto C, Nakajima E, Isagai T, Matsuishi T (2003) Possible association between congenital cytomegalovirus infection and autistic disorder. J Autism Dev Disord 33(4):455–459

Yurchenko E, Shio MT, Huang TC, Da Silva Martins M, Szyf M, Levings MK, Olivier M, Piccirillo CA (2012) Inflammation-driven reprogramming of CD4+ Foxp3 + regulatory T cells into pathogenic Th1/Th17 T effectors is abrogated by mTOR inhibition in vivo. PLoS ONE 7(4):e35572. doi:10.1371/journal.pone.0035572 PONE-D-11-25382 [pii]

Chapter 10
Antibody Mediating Autoimmune Reaction in Autism Spectrum Disorder

Elena Noris-García, Mercedes Adalys Rodríguez-Ravelo, Yamila Adams Villalón, Gustavo Sierra and Maria de los Angeles Robinson-Agramonte

Abstract Autism spectrum disorders (ASD) are a group of severe pervasive neurodevelopmental disorders characterized by a highly variable impairment in social interaction and verbal and nonverbal communication, a stereotyped repetitive behavior, learning problems, and aloofness. ASD is frequently accompanied by a core neurobehavioral symptoms and immunological derangements, including aberrant sensitivity to sensory stimulation, anxiety and cellular immune deregulation. Currently, the involvement of the neuroimmune pathology in autism remains unclear and we consider that a better understanding would be useful for earlier clinical and therapeutic intervention. The main aim of this chapter is to review the most current aspects regarding the antibody response in autism occurring either in the periphery or into the brain and how they can influence the abnormal development of the offspring and modulate the nontypical behaviors frequently observed in autism.

E. Noris-García (✉)
Psychoneuroimmunology Department, Nephrology Institute, Medical Havana University,
Ave 26, Esq. Rancho Boyeros, Havana 11600, Cuba
e-mail: anoris@infomed.sld.cu

M. A. Rodríguez-Ravelo
Endocrinology Laboratory Ameijeiras Brothers Hospital, Medical Havana University,
San Lázaro 706 esquina Belascoin, Havana, Cuba
e-mail: rodriguezadalys@yahoo.com

Y. A. Villalón
Institute of Hematology and Immunology, Calzada de Aldabó y Calle E, Altahabana, CP 10300
La Habana, Cuba
e-mail: yamila.adams@infomed.sld.cu

M. d. l. A. Robinson-Agramonte
Neuroimmunology Department, International Center for Neurological
Restoration, Havana, Cuba
e-mail: maria.robinson@infomed.sld.cu

G. Sierra
BioCubaFarma: Group of Biotechnological and Pharmaceutical Industries of Cuba,
Ave Independencia No. 8126 Esq. Calle 100, Boyeros, Havana, Cuba
e-mail: gsierra@oc.biocubafarma.cu

M. d. l. A. Robinson-Agramonte (ed.), *Translational Approaches to Autism Spectrum Disorder*, DOI 10.1007/978-3-319-16321-5_10

Keywords Neurimmunology · ASD · Neurodevelopmental disorders · Humoral immunity · Autoantibody · Autoimmunity

10.1 Introduction

Certain autoimmune disorders have a robust impact behavior. This has been evidenced among people with systemic diseases showing neuropsychiatric symptoms (Diamond et al. 2006) With a four times higher incidence in boys than in girls, autism is accompanied by core symptoms of neurobehavioral and immunological derangements, including humoral immunity response antibody mediated. This chapter is aimed to describe current aspects of humoral immunity in autism interacting within and outside the brain.

10.2 Autism's Etiological Hypothesis

Several studies have provided evidence suggesting that autism is a neurodevelopmental disorder with epigenetic factors, where a pathophysiological process arises from the interaction of an early environmental insult with the genetic predisposition inherited. (Hallmayer et al. 2011). Prenatal and early postnatal factors involve elements from humoral arm of the immune response in autism (Singer et al. 2009; Niehus and Lord 2006; Hallmayer et al. 2011)(Vojdani et al. 2003; Ashwood and Wakefield 2006; Niehus and Lord 2006; Rosen et al. 2007). Some evidences associated with increased risk for ASD are referred as consequences of autoimmune phenomena around pregnancy (Ashwood et al. 2006; Cohly and Panja 2005), or by interaction between maternal or fetal environmental, at prenatal or perinatal periods (Fombonne 2005a, 2005b; Hertz-Picciotto et al. 2006) generating a global immune dysregulation during gestation.

In line with this idea, an aberrant immune function in individuals affected by ASD has been reported for nearly 40 years with a long-standing focus on autoimmune elements (Cooper 2003; Hertz-Picciotto et al. 2006) at least, in a subgroup of patients with autism, and following either the autoimmune reaction generated by the global immune deregulation during gestation or by the abnormal interaction between maternal or fetal environmental and genetic factors, occurring at prenatal or perinatal periods (Fombonne 2005a, 2005b; Hertz-Picciotto et al. 2006).

10.3 Immunoglobulins Imbalance in ASD

Immunoglobulin (Ig) G is the most prevalent antibody isotype in human circulation with four subclasses, each one with different biological properties and function: IgG1 and IgG3, predominately responsible for protection against reinfection based mainly on their ability to complement activation and the IgG2 and IgG4 a few less protagonist, nevertheless, none clear clinical implications of IgG subclasses has been established.

In case of children with autism an increased serum IgG2 and IgG4 concentrations were demonstrated, associated with certain behavioral outcomes (Croonenberghs et al. 2002b, 2002a). A case-control study demonstrated increased IgG4 level in children with autism, compared to developmentally delayed children unaffected by autism (Enstrom et al. 2010). On the contrary, one study found an negative relationship between IgG/IgM levels and behavioral symptoms: those patients with the highest scores in the behavioral battery had the most reduced levels of IgG and IgM, and the diminished IgM levels were more pronounced within the autistic group showing the most regressive phenotype (Heuer et al. 2008).

While these reports appear to be contradictory, different methodological and epidemiological factors could underlie these results: the use of small sample sizes (ranging between 18 and 40 autistic subjects); the type of controls used in the comparisons (siblings vs. matched general population vs. patients with other neuropsyquiatric diseases); the age of the patients (Ig levels do not reach adult levels until 1 year of age for IgM, age 6–8 years for IgG, and age 10 years for IgA, so comparisons within a broad age range, especially during the age spanning immune development, may introduce artifacts); geographic location of the subjects and controls (at tropical regions, for example, people in general are more susceptible to variations of Ig levels); or, the season during which the sample was obtained (this determines, for example, the exposure to immunogens and allergens).

10.4 Antibodies Contributing to Brain Damage in ASD

Regardless of the actual differences in the total amounts of different Igs, the specificity of these antibodies can be more relevant to understand the etiology and/or pathophysiology of autism. Reports of self-reactive antibodies, especially to brain and central nervous system (CNS) proteins, have been considered as one of the most robust and consistently immunological findings in autism (Cabanlit et al. 2007; Singer et al. 2006; Singer and Williams 2006; Wills et al. 2007, 2011; Rossi et al. 2011; Al-Ayadhi and Mostafa 2011; Mostafa and Al-Ayadhi 2011b, 2011c, 2011a; Todd et al. 1988; Singh et al. 1993; Connolly et al. 1999, 2006; Vojdani et al. 2002; Singh and Rivas 2004b). It has been estimated that around 25–70 % of children with autism have antibodies reactive to neuronal proteins, indicating a potential deregulation in this network of autism (Connolly et al. 1999; Singh et al. 2009, 1997b).

The presence of autoantibodies reactive to neuronal tissue in plasma of children with autism suggests a prenatal or early postnatal etiology potentially involving an aberrant cross talk between the maternal and fetal immune systems during fetal neurodevelopment (Braunschweig et al. 2013, 2012; Nordahl et al. 2013; Fox et al. 2012; Braunschweig and Van de Water 2012; Goines et al. 2011; Croen et al. 2008a). In humans, maternal IgG isotype antibodies readily cross the placenta to equip the immunologically naïve fetus with a subset of the maternal adaptive humoral immune system proteins (Croen et al. 2008b); whereas maternal IgG antibodies which persist for up to 6 months postnatal (Braunschweig and Van de Water

2012). However, when along these protective IgGs, other antibodies reactive to fetal self-proteins cross the placenta, they can have pathological effects (Bauman et al. 2013; Nordahl et al. 2013; Braunschweig et al. 2012; Fox et al. 2012).

Autoimmune disorders associated with antibodies, such as rheumatoid arthritis, lupus, and thyroid disease, are increased among mothers and other family members of autistic children. Maternal antibodies, therefore, may influence fetal brain development during pregnancy by interfering with cell signaling in the developing brain and disturbing its organization. These antibodies also might result from environmental exposures in susceptible mothers during pregnancy.

10.4.1 Autoimmunity Mediating Prenatal Brain Damage in ASD

Studies suggesting a role for maternal antibodies in the etiology of autism reported that mothers of children with autism present serum patterns of immunoreactivity to prenatal rat brain, which are not observed in control mothers. These antibodies can be detected after delivery (Zimmerman et al. 2007) or at midpregnancy (Croen et al. 2008a). Actually, mothers of an ASD child are four times more likely to have antibodies against brain proteins than control women (Brimberg et al. 2013).

The potential role of a heightened or activated maternal immune response in the risk for ASD is further strengthened by epidemiological data from large population-based studies that show increased rates of autoimmune disorders in the families of individuals with ASD. Independently or coincidentally, the presence of specific anti-fetal brain antibodies in approximately 12 % of mothers of children with ASD, which are absent in mothers of children who are typically developing or mothers of children with developmental delays, suggests a potential inflammatory process occurring in mothers of children with ASD that leads to the production of antibodies directed to the developing brain (Onore et al. 2012; Croen et al. 2008a; Careaga and Ashwood 2012).These antibodies can be detected after delivery (Zimmerman et al. 2007) or at midpregnancy (Zimmerman et al. 2007; Croen et al. 2008a). Actually, mothers of an ASD child are four times more likely to have antibodies against brain proteins than control women (Zimmerman et al. 2007; Brimberg et al. 2013).

Others evidences suggest that the presence of maternal antibodies against fetal brain proteins of 37 and 73 kDa of molecular weight increases the risk of autism (Zimmerman et al. 2007; Braunschweig et al. 2008, 2012, 2013; Braunschweig and Van de Water 2012) or, at least, impairs verbal and nonverbal language acquisition (Piras et al. 2014). A recent study showed that sera from mothers of autistic children react specifically to proteins highly expressed during brain development (Braunschweig et al. 2013; Bauman et al. 2013). Moreover, sera from autism mothers recognize fetal brain rat proteins, but not adult ones (Zimmerman et al. 2007), suggesting a specific vulnerability during embryonic development. However, other studies characterizing the presence of autoantibodies in autistic patients suggest that they play a postnatal role (see below).

Correlational studies linking the occurrence of maternal antibodies and specific symptoms of ASD are still largely lacking. Recently, it has been shown that the presence of IgG reactive to fetal brain tissue in the maternal serum is associated with an enlarged frontal lobe in autistic children, when compared with autistic children born from mothers without these antibodies (Nordahl et al. 2013).

Although, it is not clear how these antibodies can result in the behavioral symptoms of ASD, some evidences from experimental animals show that they can affect neuronal function and alter behavior of the offspring. For example, the injection of the serum of a mother of an autistic child into pregnant mice resulted in cerebellar and behavioral alterations in the offspring. This serum contained antibodies binding to cerebellar Purkinje cells (Courchesne 1997; Fatemi et al. 2012). Similarly, the injection of purified IgGs from 63 autism mothers into pregnant mice resulted in offspring with reduced sociability (Singer et al. 2009). Finally, a single low dose gestational exposure to IgG from autism mothers results in mice showing impaired sensory and motor development and altered behavior (Braunschweig et al. 2012).

In addition, the injection of purified IgGs from 21 mothers of at least one child with autism into pregnant rhesus monkeys resulted in offspring showing atypical social behavior and increased stereotypies (Martin et al. 2008). Moreover, monkeys prenatally exposed to IgGs from autism mothers showed enlarged brain volume, especially due to increased white matter volume (Bauman et al. 2013).

10.4.2 Autoimmunity Mediating Postnatal Brain Damage in ASD

Autoantibodies directed against different CNS proteins have not only been detected in mothers, but also in autistic patients. The first report on autoantibodies against brain proteins was almost 30 years ago (Todd and Ciaranello 1985), and they were shown to bind to proteins expressed in various rat brain regions (Singh and Rivas 2004b, 2004a)(Singh and Rivas 2004b) and human brain proteins (Cabanlit et al. 2007; Singer et al. 2006, 2009). Numerous antigens have been identified, including glial and neuron-axon filament proteins (Singh et al. 1997b), myelin basic protein (MBP)(Singh and Rivas 2004b, 2004a; Libbey et al. 2008), serotonin receptors (Todd and Ciaranello 1985), nerve growth factor (Connolly et al. 2006), brain-derived neurotrophic factor (Connolly et al. 2006), and brain endothelial (Vojdani et al. 2002, 2004; Wills et al. 2007, 2011, 2009; Rout et al. 2012; Libbey et al. 2008; Kirkman et al. 2008). It is also important to note that it is not clear whether autistic children are positive for more than one antibody.

Ganglioside M1 is the most abundant ganglioside in neural membranes and specific antibodies have been detected in some autoimmune diseases with neurological involvement. Autistic patients show increased serum levels of anti-ganglioside M1 antibodies and their titer correlated with the severity of the disease (Mostafa and Al-Ayadhi 2011b; Vojdani et al. 2002, 2004; Wills et al. 2011). Nevertheless, some authors had found lack of association between these autoantibodies and autism (Moeller et al. 2013).

A study, characterizing the prevalence and specificity of autoantibodies against cerebellar tissue (Wills et al. 2009; Vojdani et al. 2002), 21 % of the ASD children presented autoantibodies reactive to a protein expressed at cerebellum, specifically by Golgi cells, while another study identified antibodies against different cerebellar proteins in autistic children (Wills et al. 2009; Vojdani et al. 2002; Goines et al. 2011), and showed that antibodies against a 45 kDa cerebellar protein are associated in an specific manner with autism,whereas, antibodies reacting to a 62 kDa cerebellar protein was associated with other disorders under the umbrella of ASD. Therefore, the specificity of the autoantibodies in ASD could underlay the different symptoms observed in autism, nevertheless, the pathological relevance of these must be clarified (Hegvik et al. 2014). Some authors reporting antibodies reactive to fetal brain proteins in autism, suggest that it may predict better behavioral and emotional problems than diagnosis (Rossi et al. 2011), while others reports claim that these antibodies do not predict autism (Morris et al. 2009) and that autoantibodies in children with autism appear to react more with the fully developed brain than with specific structure in the fetal brain unlike their mothers (Cabanlit et al. 2007; Singer 2009),as well as the previous reported immunoreactivities have been challenged (Mostafa and Al-Ayadhi 2012; Moeller et al. 2013).

Worth noticing, another study has revealed in the sera of ASD children the presence of antibodies against GABAergic neurons throughout the central nervous system, and not specifically in the cerebellum (Wills et al. 2011; Rout et al. 2012).

Moreover, the presence of neuronal protein-specific autoantibodies are associated with increased behavioral impairments and more severity in children with ASD (Mostafa and Al-Ayadhi 2011b, 2012), suggesting a link between the autoimmune processes and behavioral dysfunction. So, autoantibodies directed against a 45 kDa protein present in the cerebellum were not only found more frequently in children with ASD but were also associated with lower adaptive and cognitive function, as well as increased aberrant behaviors (Wills et al. 2011; Goines et al. 2011). However, replication studies of antibody-specific antigen targets, such as MBP and Glial Fibrillar Acid Protein (GFAP), have been inconsistent, suggesting that further studies are needed to identify the target or targets and/or associated autoimmune phenomena (Libbey et al. 2008; Kirkman et al. 2008). It would be interesting to clarify whether the presence of autoantibodies in children can be or not useful to predict ASD, and which are the pathological mechanisms resulting in the core symptoms of the disease.

10.5 Antinuclear Antibodies in ASD

Studies evaluating the occurrence of autoantibodies against nuclear proteins in autism, reported that antinuclear antibodies were present in children with autism in a high frequency (Singh and Rivas 2004a), as it is observed in other typical autoimmune diseases like systemic lupus erythematosus, but not in normal children (Colasanti et al. 2009), suggesting a link of autoimmunity and the aberrant behavioral observed in autism since neuropsychiatric symptoms are often observed in autoimmune diseases as well as serum antinuclear autoantibodies, in patients with autoimmune diseases showing neuropsychiatric symptoms, cross react with N-methyl-D-aspartate (NMDA) receptor related to glutamate neurotransmitter (Diamond and Volpe 2006; Huerta et al. 2006; Kowal et al. 2006).

Serum levels of antinucleosome-specific antibodies have been reported increased in some autistic children, especially with a family history of autoimmunity. However, their role in the induction of autoimmunity in a subgroup of autistic children is not clear (AL-Ayadhi and Mostafa 2014)

10.6 Antiphospholipid Antibodies in Autism

Antiphospholipid antibodies recognize a number of diverse targets including cardiolipin, phosphoserine, and β2-glycoprotein 1. Elevated levels of antiphospholipid antibodies have been found in the blood and cerebral spinal fluid of psychiatric patients having hallucinatory and/or delusionary episodes (Sokol et al. 2007). In individuals with neuropsychiatric systemic lupus erythematosus, elevated titers of anticardiolipin antibodies are reported most often in patients with cognitive impairment, psychosis, depression, seizures, chorea, and migraines (Blank et al. 2007; Zandman-Goddard et al. 2007). Moreover, some experiences in animal models of rodents, by administration of antiphospholipid antibodies, demonstrated the possibility to induce a number of psychological side effects, including increased anxiety and decreased cognition learning and memory (Shoenfeld et al. 1998; Lerner et al. 1998) in this pathology.

Recent studies have also found levels of anticardiolipin, β2-glycoprotein 1, and anti-phosphoserine antibodies elevated in children with ASD compared with age-matched typically developing (TD) and developmental delays (DD) controls (Careaga et al. 2013; Abisror et al. 2013). Furthermore, the increase in antibody levels was associated with more impaired behaviors reported by parents (Abisror et al. 2013; Careaga et al. 2013; Careaga and Ashwood 2012)

Autoantibodies directed toward serotonin receptors have also been demonstrated in autistic patients compared with controls (Singh and Rivas 2004b; Burgess et al. 2006; Singh et al. 1997a; Yuwiler et al. 1992; Todd and Ciaranello 1985), supporting the role of serotonin in the abnormal immune response and behavioral in autism.

These evidences argue for an increased utilization of dietary tryptophan by the gut and less tryptophan available for passage through the blood-brain barrier, which would result in reduced 5-HT levels in the brain, an additional argue to the mood and cognitive dysfunctions reported in ASD (McDougle et al. 1996; Cappiello et al. 1996). Nevertheless, from this analysis more research would be conducted to establish a potential therapeutic target in ASD to the future, either by providing dietary tryptophan or by pharmaceutical treatments based in selective serotonin reuptake inhibitors.

10.7 Environmental Factor and Autoimmune Disorders

Some potential environmental factor triggering for autoimmune disorders had been studied in autistic children. Vitamin D, for example, might play a regulatory role in the production of autoantibodies. Recent studies have found that Serum 25-hydroxy vitamin D levels had significant negative correlations with serum levels of anti-myelin-associated glycoprotein (anti-MAG) autoantibodies in some autistic children. Vitamin D deficiency may contribute to the production of these autoantibodies, through induction of specific CD4(+)CD25(high) regulatory T-cells (Mostafa and Al-Ayadhi 2012).

In relationship to autoimmune response to dietary proteins, some studies revealed high seropositivity for autoantibodies to casein and gluten might be associated with autism (Vojdani et al. 2003; Kawashti et al. 2006).

On the other hand, reduced folate transport to the CNS was identified in two autism spectrum disorders, i.e., Rett syndrome and infantile low-functioning autism with neurological abnormalities. The reduced Cerebrospinal fluid (CSF) folate in these patients could be explained by serum folate receptor (FR) autoantibodies blocking the folate binding site of the membrane-attached FR on the choroid epithelial cells. Early detection of FR autoantibodies may be a key factor in the prevention and therapeutic intervention among this subgroup of patients with autism (Vojdani et al. 2003; Ramaekers et al. 2007).

A study to see the relation between oxidative stress (by measuring plasma F2-isoprostane, as a marker of lipid peroxidation, and glutathione peroxidase, as an antioxidant enzyme) and autoimmunity (as indicated by serum antineuronal antibodies) in a group of 44 Egyptian autistic children and 44 healthy-matched children were doing.

It was found that, the strong association between oxidative stress and autoimmunity in autistic children may indicate the possible role of oxidative stress, through induction of autoimmunity, in some autistic patients (Mostafa et al. 2009)

Autoimmunity to the CNS, especially to MBP, may play a causal role in autism, a neurodevelopmental disorder. Because many autistic children harbor elevated levels of measles antibodies, a serological study of measles-mumps-rubella (MMR) and MBP autoantibodies was conducted. The high expression of MMR antibody specifically detected a protein of 73–75 kD measles hemagglutinin (HA) protein, which is unique to the measles subunit of the vaccine. Furthermore, over 90 % of MMR antibody-positive autistic sera were also positive for MBP autoantibodies, suggesting a strong association between inappropriate antibody response to MMR and CNS autoimmunity in autism (Singh et al. 2002).

However, in other research authors found that, a deficient immune response to MMR vaccine antigens might be associated with autism, as a leading cause or a resulting event (Kawashti et al. 2006).

Others studies demonstrated that dietary peptides, bacterial toxins, and xenobiotics bind to lymphocyte receptors and/or tissue enzymes, resulting in autoimmune reaction with autoantibody production, in children with autism (Vojdani et al. 2003).

10.8 Conclusions

Although it is well known that autism affects primarily the brain function (mainly social functioning and cognition), it is also known that it follows the affectation of other organs and systems. So, published evidences have identified widespread changes in the humoral immune response of children with autism and neuroimmuno-endocrine interaction. Brain specimens from autistic subject's exhibit signs of active, ongoing inflammation. These reports suggest that autism may in fact be a systemic disorder where under the influence of prenatal and around birth factors, an early brain development alter is caused, which is potentially critical to induce autism development.

References

Abisror N, Mekinian A, Lachassinne E, Nicaise-Roland P, De PL, Chollet-Martin S, Boddaert N, Carbillon L, Fain O (2013) Autism spectrum disorders in babies born to mothers with antiphospholipid syndrome. Semin Arthr Rheum 43:348–351

Al-Ayadhi LY, Mostafa GA (2011) Increased serum osteopontin levels in autistic children: relation to the disease severity. Brain Behav Immun 25:1393–1398

Al-Ayadhi LY, Mostafa GA (2014) Serum antinucleosome-specific antibody as a marker of autoimmunity in children with autism. J Neuroinflammation 11:69

Ashwood P, Wakefield AJ (2006) Immune activation of peripheral blood and mucosal CD3 + lymphocyte cytokine profiles in children with autism and gastrointestinal symptoms. J Neuroimmunol 173(1–2):126–134

Ashwood P, Wills S, Van De Water J (2006) The immune response in autism: a new frontier for autism research. J Leukoc Biol 80:1–15

Bauman MD, Iosif AM, Ashwood P, Braunschweig D, Lee A, Schumann CM, Van De Water J, Amaral DG (2013) Maternal antibodies from mothers of children with autism alter brain growth and social behavior development in the rhesus monkey. Transl Psychiatry 3:e278. doi:10.1038/tp.2013.47

Blank M, Anafi L, Zandman-Goddard G, Krause I, Goldman S, Shalev E, Cervera R, Font J, Fridkin M, Thiesen HJ, Shoenfeld Y (2007) The efficacy of specific IVIG anti-idiotypic antibodies in antiphospholipid syndrome (APS): trophoblast invasiveness and APS animal model. Int Immunol 19:857–865

Braunschweig D, Van De Water J (2012) Maternal autoantibodies in autism. Arch Neurol 69:693–699

Braunschweig D, Ashwood P, Krakowiak P, Hertz-Picciotto I, Hansen R, Croen LA, Pessah IN, Van De Water J (2008) Autism: maternally derived antibodies specific for fetal brain proteins. Neurotoxicology 29:226–231

Braunschweig D, Golub MS, Koenig CM, Qi L, Pessah IN, Van De Water J, Berman RF (2012) Maternal autism-associated IgG antibodies delay development and produce anxiety in a mouse gestational transfer model. J Neuroimmunol 252:56–65

Braunschweig D, Krakowiak P, Duncanson P, Boyce R, Hansen RL, Ashwood P, Hertz-Picciotto I, Pessah IN, Van De Water J (2013) Autism-specific maternal autoantibodies recognize critical proteins in developing brain. Transl Psychiatry 3:e277. doi:10.1038/tp.2013.50

Brimberg L, Sadiq A, Gregersen PK, Diamond B (2013) Brain-reactive IgG correlates with autoimmunity in mothers of a child with an autism spectrum disorder. Mol Psychiatry 18:1171–1177

Burgess NK, Sweeten TL, Mcmahon WM, Fujinami RS (2006) Hyperserotoninemia and altered immunity in autism. J Autism Dev Disord 36:697–704

Cabanlit M, Wills S, Goines P, Ashwood P, Van De Water J (2007) Brain-specific autoantibodies in the plasma of subjects with autistic spectrum disorder. Ann NY Acad Sci 1107:92–103

Cappiello A, Malison RT, Mcdougle CJ, Vegso SJ, Charney DS, Heninger GR, Price LH (1996) Seasonal variation in neuroendocrine and mood responses to i.v. L-tryptophan in depressed patients and healthy subjects. Neuropsychopharmacology 15:475–483

Careaga M, Ashwood P (2012) Autism spectrum disorders: from immunity to behavior. Methods Mol Biol 934:219–240. doi:10.1007/978-1-62703-071-712

Careaga M, Hansen RL, Hertz-Piccotto I, Van De Water J, Ashwood P (2013) Increased anti-phospholipid antibodies in autism spectrum disorders. Mediators Inflamm 2013:935608. doi:10.1155/2013/935608. (Epub@2013 Sep 23)

Cohly HH, Panja A (2005) Immunological findings in autism. Int Rev Neurobiol 71:317–341

Colasanti T, Delunardo F, Margutti P, Vacirca D, Piro E, SiracusanoA, Ortona E (2009) Autoantibodies involved in neuropsychiatric manifestations associated with systemic lupus erythematosus. J Neuroimmunol 212:3–9

Connolly AM, Chez MG, Pestronk A, Arnold ST, Mehta S, Deuel RK (1999) Serum autoantibodies to brain in Landau-Kleffner variant, autism, and other neurologic disorders. J Pediatr 134:607–613

Connolly AM, Chez M, Streif EM, Keeling RM, Golumbek PT, Kwon JM, Riviello JJ, Robinson RG, Neuman RJ, Deuel RM (2006) Brain-derived neurotrophic factor and autoantibodies to neural antigens in sera of children with autistic spectrum disorders, Landau-Kleffner syndrome, and epilepsy. Biol Psychiatry 59:354–363

Cooper EL (2003) Neuroimmunology of autism: a multifaceted hypothesis. Int J Immunopathol Pharmacol 16:289–292

Courchesne E (1997) Brainstem, cerebellar and limbic neuroanatomical abnormalities in autism. Curr Opin Neurobiol 7:269–278

Croen LA, Braunschweig D, Haapanen L, Yoshida CK, Fireman B, Grether JK, Kharrazi M, Hansen RL, Ashwood P, Van De Water J (2008a) Maternal mid-pregnancy autoantibodies to fetal brain protein: the early markers for autism study. Biol Psychiatry 64:583–588

Croen LA, Goines P, Braunschweig D, Yolken R, Yoshida CK, Grether JK, Fireman B, Kharrazi M, Hansen RL, Van De Water J (2008b) Brain-derived neurotrophic factor and autism: maternal and infant peripheral blood levels in the Early Markers for Autism (EMA) Study. Autism Res 1:130–137

Croonenberghs J, Bosmans E, Deboutte D, Kenis G, Maes M (2002a) Activation of the inflammatory response system in autism. Neuropsychobiology 45:1–6

Croonenberghs J, Wauters A, Devreese K, Verkerk R, Scharpe S, Bosmans E, Egyed B, Deboutte D, Maes M (2002b) Increased serum albumin, gamma globulin, immunoglobulin IgG, and IgG2 and IgG4 in autism. Psychol Med 32:1457–1463

Diamond B, Volpe BT (2006) Antibodies and brain disease: a convergence of immunology and physiology. PLoS Med 3:e498

Diamond B, Kowal C, Huerta PT, Aranow C, Mackay M, Degiorgio LA, Lee J, Triantafyllopoulou A, Cohen-Solal J, Volpe BT (2006) Immunity and acquired alterations in cognition and emotion: lessons from SLE. Adv Immunol 89:289–320

Enstrom AM, Onore CE, Van De Water JA, Ashwood P (2010) Differential monocyte responses to TLR ligands in children with autism spectrum disorders. Brain Behav Immun 24:64–71

Fatemi SH, Aldinger KA, Ashwood P, Bauman ML, Blaha CD, Blatt GJ, Chauhan A, Chauhan V, Dager SR, Dickson PE, Estes AM, Goldowitz D, Heck DH, Kemper TL, King BH, Martin LA, Millen KJ, Mittleman G, Mosconi MW, Persico AM, Sweeney JA, Webb SJ, Welsh JP (2012) Consensus paper: pathological role of the cerebellum in autism. Cerebellum 11:777–807

Fombonne E (2005a) Developing modern child psychiatry. World Psychiatry 4:156–157

Fombonne E (2005b) Epidemiology of autistic disorder and other pervasive developmental disorders. J Clin Psychiatry 66(Suppl 10):3–8

Fox E, Amaral D, Van De Water J (2012) Maternal and fetal antibrain antibodies in development and disease. Dev Neurobiol 72:1327–1334

Goines P, Haapanen L, Boyce R, Duncanson P, Braunschweig D, Delwiche L, Hansen R, Hertz-Picciotto I, Ashwood P, Van De Water J (2011) Autoantibodies to cerebellum in children with autism associate with behavior. Brain Behav Immun 25:514–523

Hallmayer J, Cleveland S, Torres A, Phillips J, Cohen B, Torigoe T, Miller J, Fedele A, Collins J, Smith K, Lotspeich L, Croen LA, Ozonoff S, Lajonchere C, Grether JK, Risch N (2011) Genetic heritability and shared environmental factors among twin pairs with autism. Arch Gen Psychiatry 68:1095–1102

Hegvik TA, Husebye ES, Haavik J (2014) Autoantibodies targeting neurotransmitter biosynthetic enzymes in attention-deficit/hyperactivity disorder (ADHD). Eur Child Adolesc Psychiatry 23:115–117

Hertz-Picciotto I, Croen LA, Hansen R, Jones CR, Van De Water J, Pessah IN (2006) The CHARGE study: an epidemiologic investigation of genetic and environmental factors contributing to autism. Environ Health Perspect 114:1119–1125

Heuer L, Ashwood P, Schauer J, Goines P, Krakowiak P, Hertz-Picciotto I, Hansen R, Croen LA, Pessah IN, Van De Water J (2008) Reduced levels of immunoglobulin in children with autism correlates with behavioral symptoms. Autism Res 1:275–283

Huerta PT, Kowal C, Degiorgio LA, Volpe BT, Diamond B (2006) Immunity and behavior: antibodies alter emotion. Proc Natl Acad Sci U S A 103:678–683

Kawashti MI, Amin OR et al (2006) Possible immunological disorders in autism: concomitant autoimmunity and immune tolerance. Egypt J Immunol 13(1):99–104

Kirkman NJ, Libbey JE, Sweeten TL, Coon HH, Miller JN, Stevenson EK, Lainhart JE, Mcmahon WM, Fujinami RS (2008) How relevant are GFAP autoantibodies in autism and Tourette Syndrome? J Autism Dev Disord 38:333–341

Kowal C, Degiorgio LA, Lee JY, Edgar MA, Huerta PT, Volpe BT, Diamond B (2006) Human lupus autoantibodies against NMDA receptors mediate cognitive impairment. Proc Natl Acad Sci U S A 103:19854–19859

Lerner A, Blank M, Lahat N, Shoenfeld Y (1998) Increased prevalence of autoantibodies in celiac disease. Dig Dis Sci 43:723–726

Libbey JE, Coon HH, Kirkman NJ, Sweeten TL, Miller JN, Stevenson EK, Lainhart JE, Mcmahon WM, Fujinami RS (2008) Are there enhanced MBP autoantibodies in autism? J Autism Dev Disord 38:324–332

Martin LA, Ashwood P, Braunschweig D, Cabanlit M, Van De Water J, Amaral DG (2008) Stereotypies and hyperactivity in rhesus monkeys exposed to IgG from mothers of children with autism. Brain Behav Immun 22:806–816

McDougle CJ, Naylor ST, Cohen DJ, Aghajanian GK, Heninger GR, Price LH (1996) Effects of tryptophan depletion in drug-free adults with autistic disorder. Arch Gen Psychiatry 53:993–1000

Moeller S, Lau NM, Green PH, Hellberg D, Higgins JJ, Rajadhyaksha AM, Alaedini A (2013) Lack of association between autism and anti-GM1 ganglioside antibody. Neurology 81:1640–1641

Morris CM, Pardo-Villamizar C, Gause CD, Singer HS (2009) Serum autoantibodies measured by immunofluorescence confirm a failure to differentiate PANDAS and Tourette syndrome from controls. J Neurol Sci 276:45–48

Mostafa GA, Al-Ayadhi LY (2011a) A lack of association between hyperserotonemia and the increased frequency of serum anti-myelin basic protein auto-antibodies in autistic children. J Neuroinflammation 8:71–78. doi:10.1186/1742-2094-8-71

Mostafa GA, Al-Ayadhi LY (2011b) Increased serum levels of anti-ganglioside M1 auto-antibodies in autistic children: relation to the disease severity. J Neuroinflammation 8:39. doi:10.1186/1742-2094-8-39

Mostafa GA, Al-Ayadhi LY (2011c) The possible link between the elevated serum levels of neurokinin A and anti-ribosomal P protein antibodies in children with autism. J Neuroinflammation 8:180–188. doi:10.1186/1742-2094-8-180

Mostafa GA, Al-Ayadhi LY (2012) The relationship between the increased frequency of serum antineuronal antibodies and the severity of autism in children. Eur J Paediatr Neurol 16:464–468
Mostafa GA, El-Hadidi ES et al (2009) Oxidative stress in Egyptian children with autism: relation to autoimmunity. J Neuroimmunol 219(1–2):114–118
Niehus R, Lord C (2006) Early medical history of children with autism spectrum disorders. J Dev Behav Pediatr 27:S120–S127
Nordahl CW, Braunschweig D, Iosif AM, Lee A, Rogers S, Ashwood P, Amaral DG, Van De Water J (2013) Maternal autoantibodies are associated with abnormal brain enlargement in a subgroup of children with autism spectrum disorder. Brain Behav Immun 30:61–65. doi:10.1016/j.bbi.2013.01.084. (Epub@2013 Feb 8)
Onore C, Careaga M, Ashwood P (2012) The role of immune dysfunction in the pathophysiology of autism. Brain Behav Immun 26:383–392
Piras IS, Haapanen L, Napolioni V, Sacco R, Van De Water J, Persico AM (2014) Anti-brain antibodies are associated with more severe cognitive and behavioral profiles in Italian children with Autism Spectrum Disorder. Brain Behav Immun 38:91–99. doi:10.1016/j.bbi.2013.12.020. (Epub@2014 Jan 3)
Ramaekers VT, Blau N et al (2007) Folate receptor autoimmunity and cerebral folate deficiency in low-functioning autism with neurological deficits. Neuropediatrics 38(6):276–281
Rosen NJ, Yoshida CK et al (2007) Infection in the first 2 years of life and autism spectrum disorders. Pediatrics 119(1):e61–e69
Rossi CC, Van De Water J, Rogers SJ, Amaral DG (2011) Detection of plasma autoantibodies to brain tissue in young children with and without autism spectrum disorders. Brain Behav Immun 25:1123–1135
Rout UK, Mungan NK, Dhossche DM (2012) Presence of GAD65 autoantibodies in the serum of children with autism or ADHD. Eur Child Adolesc Psychiatry 21:141–147
Shoenfeld Y, Sherer Y, Blank M (1998) Antiphospholipid syndrome in pregnancy–animal models and clinical implications. Scand J Rheumatol Suppl 107:33–36
Singer HS (2009) Motor stereotypies. Semin Pediatr Neurol 16:77–81
Singer HS, Williams PN (2006) Autoimmunity and pediatric movement disorders. Adv Neurol 99:166–178
Singer HS, Morris CM, Williams PN, Yoon DY, Hong JJ, Zimmerman AW (2006) Antibrain antibodies in children with autism and their unaffected siblings. J Neuroimmunol 178:149–155
Singer HS, Morris C, Gause C, Pollard M, Zimmerman AW, Pletnikov M (2009) Prenatal exposure to antibodies from mothers of children with autism produces neurobehavioral alterations: a pregnant dam mouse model. J Neuroimmunol 211:39–48
Singh VK, Rivas WH (2004a) Detection of antinuclear and antilaminin antibodies in autistic children who received thimerosal-containing vaccines. J Biomed Sci 11:607–610
Singh VK, Rivas WH (2004b) Prevalence of serum antibodies to caudate nucleus in autistic children. Neurosci Lett 355(1–2):53–56
Singh VK, Warren RP et al (1993) Antibodies to myelin basic protein in children with autistic behavior. Brain Behav Immun 7(1):97–103
Singh VK, Singh EA, Warren RP (1997a) Hyperserotoninemia and serotonin receptor antibodies in children with autism but not mental retardation. Biol Psychiatry 41(6):753–755
Singh VK, Warren R, Averett R, Ghaziuddin M (1997b) Circulating autoantibodies to neuronal and glial filament proteins in autism. Pediatr Neurol 17:88–90
Singh VK, Lin SX et al (2002) Abnormal measles-mumps-rubella antibodies and CNS autoimmunity in children with autism. J Biomed Sci 9(4):359–364
Singh J, Illes J, Lazzeroni L, Hallmayer J (2009) Trends in US autism research funding. J Autism Dev Disord 39:788–795
Sokol DK, O'brien RS, Wagenknecht DR, Rao T, Mcintyre JA (2007) Antiphospholipid antibodies in blood and cerebrospinal fluids of patients with psychosis. J Neuroimmunol 190:151–156
Todd RD, Ciaranello RD (1985) Demonstration of inter- and intraspecies differences in serotonin binding sites by antibodies from an autistic child. Proc Natl Acad Sci U S A 82:612–616

Todd RD, Hickok JM et al (1988) Antibrain antibodies in infantile autism. Biol Psychiatry 23(6):644–647

Vojdani A, Campbell AW, Anyanwu E, Kashanian A, Bock K, Vojdani E (2002) Antibodies to neuron-specific antigens in children with autism: possible cross-reaction with encephalitogenic proteins from milk, Chlamydia pneumoniae and Streptococcus group A. J Neuroimmunol 129:168–177

Vojdani A, Pangborn JB et al (2003) Infections, toxic chemicals and dietary peptides binding to lymphocyte receptors and tissue enzymes are major instigators of autoimmunity in autism. Int J Immunopathol Pharmacol 16(3):189–199

Vojdani A, O'bryan T, Green JA, Mccandless J, Woeller KN, Vojdani E, Nourian AA, Cooper EL (2004) Immune response to dietary proteins, gliadin and cerebellar peptides in children with autism. Nutr Neurosci 7:151–161

Wills S, Cabanlit M, Bennett J, Ashwood P, Amaral D, Van De Water J (2007) Autoantibodies in autism spectrum disorders (ASD). Ann NY Acad Sci 1107:79–91

Wills S, Cabanlit M, Bennett J, Ashwood P, Amaral DG, Van De Water J (2009) Detection of autoantibodies to neural cells of the cerebellum in the plasma of subjects with autism spectrum disorders. Brain Behav Immun 23:64–74

Wills S, Rossi CC, Bennett J, Martinez CV, Ashwood P, Amaral DG, Van De Water J (2011) Further characterization of autoantibodies to GABAergic neurons in the central nervous system produced by a subset of children with autism. Mol Autism 2:5. doi:10.1186/2040-2392-2-5

Yuwiler A, Shih JC, Chen CH, Ritvo ER, Hanna G, Ellison GW, King BH (1992) Hyperserotoninemia and antiserotonin antibodies in autism and other disorders. J Autism Dev Disord 22:33–45

Zandman-Goddard G, Chapman J, Shoenfeld Y (2007) Autoantibodies involved in neuropsychiatric SLE and antiphospholipid syndrome. Semin Arthritis Rheum 36:297–315

Zimmerman AW, Connors SL, Matteson KJ, Lee LC, Singer HS, Castaneda JA, Pearce DA (2007) Maternal antibrain antibodies in autism. Brain Behav Immun 21:351–357

Chapter 11
Current Therapies

Dario Siniscalco and Nicola Antonucci

Abstract Nowadays, autism could be considered as a still untreatable disease. Its very complex nature and the still unclear pathophysiology and pathogenesis create challenges in treating this disorder. Current drug options for autism management are not effective for the treatment of core symptoms of autism.

This chapter describes the state-of-the-art about pharmacological interventions and focuses on novel potential therapeutics, including cell therapies, for autism care.

Keywords Autism · Pharmacotherapy · Cell therapy

11.1 Introduction

11.1.1 Autism: A Brief Overview

Autism and autism spectrum disorders (ASDs) are complex heterogeneous neurodevelopmental conditions (American Psychiatric Association 2000). Interaction of genetic and environmental factors contributes to the development of these multifactorial disorders (Siniscalco et al. 2013a). Dysfunctions in communication skills and social interactions, combined with repetitive, restrictive, and stereotypic verbal

D. Siniscalco (✉)
Department of Experimental Medicine—Division of Pharmacology,
Second University of Naples, via S. Maria di Costantinopoli, Caserta, Italy
e-mail: dariosin@uab.edu

N. Antonucci
Biomedical Centre for Autism Research and Treatment, Bari, Italy

M. d. l. A. Robinson-Agramonte (ed.), *Translational Approaches to Autism Spectrum Disorder*, DOI 10.1007/978-3-319-16321-5_11

and nonverbal behaviors are the common symptoms. Despite extensive research efforts, the etiologic-pathologies of ASDs are still inadequately understood (Siniscalco 2013; Siniscalco and Cubala-Kucharska 2013; Siniscalco et al. 2012b).

Estimated ASD prevalence is still raising; these disorders are now recognized as urgent public health problems, as their frequency continues to increase dramatically, until present dramatic rates of 1 in 68 children aged 8 years in the USA, according to Center for Disease Control (Baio 2014).

In addition, ASDs show a negative impact on the quality of life of patients and their families (Siniscalco et al. 2013b). Increased stress and mental and physical health problems have been reported in parents of autistic children compared with parents of typically developing children (Karst and Van Hecke 2012).

Health system has also been affected, as the estimated total lifetime societal cost of caring one autistic patient is US $ 3.2 million (Siniscalco 2013). Indeed, autism is a pathology that persists throughout the entire existence, hence autistic individual needs, throughout their life, continuous protection, comprehensive and accessible health care services, and prolonged specialized assistance and opportunities for independent adult life by the family (Siniscalco et al. 2013b; Zablotsky et al. 2014).

For a molecular point of view, several cellular and biochemical events are associated with ASDs: cellular oxidative stress; endoplasmic reticulum stress; decreased methylation capacity; limited production of glutathione; mitochondrial dysfunction; intestinal dysbiosis; increased toxic metal burden; immune dysregulation; immune activation of neuroglial cells (Siniscalco et al. 2012b).

The very complex nature of autism and the still unclear pathophysiology and pathogenesis create challenges in treating these disorders. However, there is no effective pharmacological intervention for the treatment of core symptoms of ASDs (Siniscalco and Antonucci 2013a)].

11.2 Autism: Current Therapies

First of all, it must be pointed out that many autism clinician practitioners agree that an early, intensive, individual-designed, and multitreatment program will greatly improve the condition of most young autistic children. Several treatments could be enrolled; however, a combination of techniques is to be preferred. Current available treatments for autism can be divided into: behavioral, nutritional, psychotherapeutical, and pharmacological approaches—even if a defined standard approach does not exists. The most used approaches include: psychological intervention, occupational and physical therapy, speech-language therapy, medications and sensory integration, and vision therapy (http://umm.edu/health/medical/ency/articles/autism Siniscalco and Antonucci 2013b).

Nowadays, it is well recognized that once a child is diagnosed with autism, the psychological intervention is a priority (Siniscalco et al. 2013b). The widely used and the most effective psychological treatment is the applied behavior analysis

(ABA). This program uses a one-on-one teaching approach useful for reinforcing the practice of various skills and is usually performed in the child's home under the supervision of a behavioral psychologist. With the use of this program, psychologists try to reach the goal of getting the child close to normal developmental functioning.

Another used program is the Treatment and Education of Autistic and Related Communication Handicapped Children (TEACCH). TEACCH uses picture schedules and other visual cues that help the child work independently, and organize and structure their environments. Unlike ABA programs, TEACCH programs try to improve child's adaptation and skills, and also accept the problems associated with autism without trying to overcome them.

11.2.1 Drugs

Pharmacological treatments only target secondary neuropsychiatric symptoms (i.e., irritability, depression, anxiety, and obsessive-compulsive behaviors) coassociated with ASDs, without addressing the core symptoms (social interaction and communication deficits, restricted and repetitive behaviors) (Farmer et al. 2013). Several drugs are prescribed: psychostimulants, alpha-2 agonists, beta blockers, lithium, anticonvulsant mood stabilizers, atypical antipsychotics, traditional antipsychotics, selective serotonin reuptake inhibitors (SSRIs), antidepressants, and antipsychotics. Developing new drugs for autism has been challenging because of a limited understanding of its pathophysiology, difficulties in establishing efficacious models of the disease and the heterogeneity of symptoms (Ghosh et al. 2013). However, in the past years several drugs have been evaluated and due the lack of efficacy have been eliminated as possible treatments for the core symptoms of autism. In example, the indirect serotonin (5-hydroxytryptamine) agonist fenfluramine (a drug approved in the USA to treat obesity) has been proposed for treating autism, due to its potential to decrease blood levels of serotonin (Cook and Leventhal 1996). The serotonin system is involved in autism and many autistic patients show elevated levels in blood. However, no therapeutic effect of fenfluramine on autistic core symptoms was achieved. In addition, several adverse effects on learning and, above all, warnings on associated cardiac valvular disease led to the removal of fenfluramine from the market in 1997 (Farmer et al. 2013).

Anticonvulsant mood stabilizers were frequently used, likely to treat epilepsy-associated symptoms (Farmer et al. 2013). Several clinical studies, however, demonstrated the lack of effects in treating autism core symptoms for divalproex sodium, lamotrigine, and levetiracetam (Hollander et al. 2010; Belsito et al. 2001; Wasserman et al. 2006). Although valproate sodium (VPA) and other mood stabilizers were considered safe and well-tolerated and could be effective for the coassociated symptoms, such as irritability and aggression, as result of the studies, this class of drugs does not treat the autism core symptoms. In addition, it is very important to consider that, on the other hand, VPA is used as autism triggering drug in animal models (Patterson 2011). VPA is also a potent neurobehavioral and neurocognitive teratogen for humans and should be strictly avoided during pregnancy (Martin and Manzoni 2014; Gentile 2014).

Among the atypical antipsychotic drugs, the most prescribed for autism is risperidone (Systematic IUPAC name 4-[2-[4-(6-fluorobenzo[*d*]isoxazol-3-yl)-1-piperidyl]ethyl]-3-methyl-2,6-diazabicyclo[4.4.0]deca-1,3-dien-5-one). Early approved by US Food and Drug Administration (FDA) for schizophrenia, it is now used to treat irritability, violent meltdowns, tantrums, aggression, and self-injury in autistic patients. Together with aripiprazole, it is the unique atypical antipsychotic drug approved by FDA for using in children and adolescents (McKinney and Renk 2011). Risperidone was also approved by the European Medicines Agency for use in ASDs. Risperidone is a selective postsynaptic monoaminergic antagonist with high affinity for serotonin 5-HT2A and dopamine D2 family receptors, as well as the D1 receptors. Adverse effects are impressive, whereas long-term effects are not clear. The most commonly reported included some extrapyramidal symptoms, somnolence, fatigue, anxiety, increased prolactin, upper respiratory tract infection, tremor, dystonia, hyperglycaemia, drowsiness, nasal congestion, increased saliva, dry mouth, enuresis, diarrhea/constipation, weight gain, and increased appetite (Baribeau and Anagnostou 2014). The increase in weight could be linked to its antagonistic actions also at the 5-HT2C receptor. Risperidone is also able to bind α1 (hypotensive effects and sedating effects) and α2 adrenergic receptors and histamine H1 receptors (sedation and reduction in vigilance) (Brunton et al. 2010). While the short-term benefit of risperidone in ameliorating severe disruptive behavior, such as irritability, in autistic children is well established (Troost et al. 2005), suggesting its use at low dose as emerging drug for the treatment of irritability associated with autistic disorder in children and adolescents (Scott and Dhillon 2008); caution should be given for its adverse effects and unknown long-term effects. Several clinicians believe that the ratio benefits/side effects is not enough to justify this medication in ASDs management.

Paliperidone, an active metabolite of risperidone, approved for schizophrenia and schizoaffective disorders in the USA, was able to decrease the irritability associated with autism in a prospective, 8-week open-label small study conducted in 25 adolescents and young adults with autism (Stigler et al. 2012).

Aripiprazole (IUPAC name 7-[4-[4-(2,3-dichlorophenyl)piperazin-1-yl]butoxy]-3,4-dihydro-1H-quinolin-2-one), a partial dopamine agonist, is an atypical antipsychotic also used to treat irritability and hyperactivity in ASDs (Ching and Pringsheim 2012). It is sometimes included in third-generation antipsychotic class, since it is aripiprazole, is a partial agonist at both dopamine D2 and serotonin 5-HT1A receptors and an antagonist at serotonin 5-HT2A, whereas second-generation antipsychotics show several levels of antagonism at the D2 receptors (Farmer and Aman 2011). Its pharmacodynamics is very particular. Indeed, aripiprazole works as agonist or antagonist in correlation to receptor density and local concentrations of dopamine. It antagonizes the D2 receptor under hyperdopaminergic conditions, and is an agonist under hypodopaminergic conditions (Kikuchi et al. 1995). It is also a 5-HT2C partial agonist and an inverse agonist at the 5-HT2B receptor. Also for this drug, major adverse effects include weight gain, sleepiness, sedation, drooling, and tremors and long-term safety and effects are unknown (Ching and Pringsheim 2012). However, some authors reported that there is no weight gain and no significant changes in glucose or lipid metabolism with its use, indicating a significant advantage over risperidone (Greenaway and Elbe 2009).

The results of clinical trials strongly suggest that the effects of atypical antipsychotics on the core symptoms of autism are very modest. Sometimes, the perceived improvements in the core symptoms are just related to the improvement in irritability and no study has been conducted to evaluate the core symptoms as a primary outcome (Farmer and Aman 2011). As stated above, the ratio benefits/side effects associated with the atypical antipsychotics does not justify their use and reconsidering the use of atypical antipsychotic in autism care is strongly encouraged (McKinney and Renk 2011).

The SSRIs are the most commonly prescribed off-label antidepressants for autism. Also for these drugs, their use is for the treatment of autism comorbidity, such as depression, anxiety, and obsessive-compulsive behaviors (Williams et al. 2013). A huge double-blind, placebo-controlled study on the use of citalopram (IUPAC name (RS)-1-[3-(dimethylamino)propyl]-1-(4-fluorophenyl)-1,3-dihydroisobenzofuran-5-carbonitrile) for autism core symptoms showed no effect of citalopram treatment over placebo. On this basis, authors concluded that the results do not support the use of citalopram for the treatment of repetitive behavior in ASD children and adolescent (King et al. 2009). Clomipramine (IUPAC name 3-(3-chloro-10,11-dihydro-5H-dibenzo[b, f]azepin-5-yl)-N, N-dimethylpropan-1-amine) a tricyclic antidepressant showed no effects on stereotypy (Remington 2001). The results of several randomized clinical trials agree that the efficacy of SSRIs for the treatment of repetitive behaviors in ASDs is modest and whether or not these drugs are effective is still unclear (Carrasco et al. 2012). Tolerability and appropriate dose need further studies. In addition, concerns about the association between the use of SSRIs during pregnancy and an increased risk of autism spectrum disorders in the offspring have been raised and need to be evaluated (Hviid et al. 2013; Rai et al. 2013).

11.2.2 Other Drugs

Cholinergic drugs, such as acetylcholinesterase inhibitors used for dementia treatment, have been proposed to treat autism, based on the fact that nicotinic receptors and M1 muscarinic receptors were found decreased in brain specimens from autistic individuals. Rivastigmine tartrate (IUPAC name (S)-3-[1-(dimethylamino)ethyl]phenyl N-ethyl-N-methylcarbamate), a cholinesterase inhibitor (it inhibits both butyrylcholinesterase and acetylcholinesterase), showed positive effects in ameliorating speech and overall autistic behavior in a 12-week open-label clinical trial on 32 autistic patients (Chez et al. 2004). However, the anticholinesterase tacrine (IUPAC name 1,2,3,4-tetrahydroacridin-9-amine) showed modest effects in the short-term treatment of irritability in autistic children. Hepatotoxicity makes it not recommended as a helpful drug for autism (Niederhofer 2007). Donepezil (IUPAC name (RS)-2-[(1-benzyl-4-piperidyl)methyl]-5,6-dimethoxy-2,3-dihydroinden-1-one) decreased irritability and hyperactivity in a small study. No changes in the core autistic symptoms were reported; side effects were gastrointestinal disturbances and mild irritability (Hardan and Handen 2002). In a later study, donezepil was effective in decreasing rapid eye movement sleep in autistic children; how this sleep state is correlated to cognition and behavior in autism remains to be elucidated (Buckley et al. 2011). Aggression and inattention were also decreased in an open-label study on the use of galantamine (IUPAC name (4aS,6R,8aS)-5,6,9,10,11,12-hexahydro-3-methoxy-11-methyl-4aH- [1]benzofuro[3a,3,2-ef] [2] benzazepin-6-ol) (Nicolson 2006). Safety alerts were released by FDA for its use in mild cognitive impairment after several deaths related to vascular causes (http://www.fda.gov/Safety/MedWatch/SafetyInformation/SafetyAlertsforHumanMedicalProducts/ucm152595.htm). Lastly, mecamylamine (IUPAC name (1S,2R,4R)-N,2,3,3-tetramethylbicyclo[2.2.1]heptan-2-amine), a nonselective and noncompetitive nicotinic receptor blocker, was ineffective in a placebo-controlled pilot trial (Arnold et al. 2012).

It has been proposed a role for glutamate excitotoxicity in the pathogenesis of autism (Essa et al. 2013). Language functions and social behavior were improved with the use of memantine (IUPAC name 3,5-dimethyltricyclo[3.3.1.1^{3,7}]decan-1amine), a nonselective NMDA receptor blocker, but also a dopamine D2 receptor agonist (Seeman et al. 2008), in autistic children over a 21-month period (Chez et al. 2007). No serious side effects were reported. However, an open-label pilot study reported only moderate improvements in cognitive functioning and behavioral symptoms (Owley et al. 2006).

Taken together, these results highlight that the efficacious and definitive pharmacologic treatment for the core symptoms of autism does not exist (Farmer et al. 2013). Hence, the lack of effective drugs have led to an increased sense of urgency to identify novel targets and therapies for ASDs (Ghosh et al. 2013).

11.2.3 Autism: Other Therapies

It has been suggested a therapeutic potential of oxytocin in treating core dimension of autism. Indeed, oxytocin, administered as nasal spray, as it does not cross the blood brain barrier (BBB), was able to improve emotion recognition and social behavior in autistic young people in respect to placebo, opening the way for its use to improve social communication and interaction in ASDs (Guastella et al. 2010). Oxytocin (IUPAC name 1-({(4R,7S,10S,13S,16S,19R)-19-amino-7-(2-amino-2-oxoethyl)-10-(3-amino-3-oxopropyl)-16-(4-hydroxybenzoyl)-13-[(1S)-1-methylpropyl]-6,9,12,15,18-pentaoxo-1,2-dithia-5,8,11,14,17-pentaazacycloicosan-4-yl}carbonyl)-L-prolyl-L-leucylglycinamide) is a hormone (nonapeptide) produced by the hypothalamus. Its action is mediated by specific G-protein-coupled receptors (oxytocin receptors). Once released by neurons, it is transported via large neurosecretory axons to the posterior hypothalamus. It has been demonstrated that changes in oxytocin receptor gene or aberrant methylation of the receptor protein, with epigenetic mechanism as protein expression regulators, are related to autism etiology (Jacob et al. 2007; Wermter et al. 2010; Insel 2010). Oxytocin-treated autistic patients exhibited more appropriate social behavior and affect (Andari et al. 2010), confirming the results obtained in rodents, suggesting that oxytocin could facilitate social information processing (Hollander et al. 2007). Further research is necessary to verify potential side effects.

Bioactive proteins and peptides could provide novel therapeutic applications for autism treatment (Siniscalco and Antonucci 2013b). Oxidative stress and mitochondrial dysfunction are among molecular changes associated with ASDs, possibly reflecting an impairment in cellular redox homeostasis. The ratio glutathione/glutathione disulfide (GSH/GSSG) is a marker of the redox-dependent cellular functions, as well as of cellular toxicity. It has been proposed that autistic children display a decreased capacity to counterbalance redox damages and detoxification capacity. Indeed, these children show decreased blood levels of total and reduced GSH, while oxidized GSH is significantly increased (Geier et al. 2009). Glutathione is a sulfhydryl tripeptide (cysteine, glycine, and glutamate) with a strong antioxidant capacity and its supplementation in autistic children, together with daily consumption of cysteine-rich foods, could increase plasma GSH levels (Kern et al. 2011; Witschi et al. 1992; McPherson and Hardy 2011).

Another interesting potential option for the management of ASD-related behaviors is the vitamin C (McGinnis 2004). This polypeptide is widely recognized as antioxidant and, together with other antioxidant peptides containing 5–16 amino acid residues, is an inhibitor of lipid peroxidation, scavengers of free radicals and chelators of transition metal ions (Siniscalco and Antonucci 2013b). Treating ASDs with antioxidant peptides could be an optimal strategy to target the oxidative stress and related autoimmunity (Mostafa et al. 2010).

11.3 Future Directions: Cellular Therapies

Nowadays, it is widely recognized that stem cell therapy represents the great promise for the future of molecular and regenerative medicine (Siniscalco et al. 2012a). In principle, stem cells provide a valid approach to curing several untreatable human diseases. Of course, stem cells represent a possibility also for curing ASDs (Siniscalco 2012). Stem cells are defined by two important and fundamental characteristics: (1) their self-renewal ability: the capacity to generate more identical stem cells; (2) differentiation process: the capacity to give rise to more differentiated cells. In addition, stem cells also show paracrine regulatory functions (Caplan and Dennis 2006): the capacity to synthesize and release a complex plethora of biomolecules (stem cells as "biopharmacy"), which in turn are able to regulate several key cellular processes, such as cell differentiation, tissue and organ repair, scar formation, apoptosis, mitosis and anti-inflammatory actions in the host recipient.

The great appeal (especially in ASDs) of stem cells, however, is due to their strong immunomodulatory capacity (Siniscalco et al. 2013c). This property is performed through the paracrine activity; by synthesizing and releasing (secretome activity) key molecules (i.e., anti-inflammatory cytokines) stem cells regulate host immune system cells (i.e., B and T cells). It is noteworthy to consider that autism is related to immune imbalance. It has been demonstrated a strong dysregulation in peripheral blood monocytic cells (PBMCs) or monocytes, the key regulators of the immune response (Siniscalco 2012c). Monocyte dysfunction could result in long-term immune alterations in autistic children (Enstrom et al. 2010). Stem cells could restore immune system also through a cell-to-cell contact activation mechanism, through which transplanted stem cells are able to switch proinflammatory macrophages to anti-inflammatory macrophages (Siniscalco et al. 2011). Through these two mechanisms stem cells could, simultaneously: counterbalance the immune system aberrations, and activate endogenous restorative mechanisms within damaged tissues contributing to recovery of functional deficits.

Among stem cell population, mesenchymal stem cells (MSCs) gained much consideration for ASD management (Siniscalco et al. 2012b). MSCs, nonhematopoietic stem cells with a multilineage potential, are easily obtained and collected from a small aspirate of bone marrow; in addition, these cells show other important characteristics useful for transplantation purposes: a high expansion potential, genetic stability, stable phenotype, high proliferation rate as adherent cells, and self-renew capacity. For their in vivo immune-modulatory properties, there are no problems with immune rejections after transplantation. There is also no tumor formation on transplantation, and no moral objection or ethical controversies are involved, as with the use of embryonic stem cells (ESCs) (Siniscalco et al. 2012b; Siniscalco et al. 2010). In addition, once transplanted, MSCs are able to migrate to the sites of injury or damage, likely driven by disease-specific chemical environment, and participate in the repair process (Li and Jiang 2011), also in this way restoring the altered brain organization is seen in autistic subjects.

In principle, other stem cell types beyond MSCs could be useful for ASD treatment (Siniscalco et al. 2013c). A very important subtype for clinical use is fetal stem cells (FSCs). These cells are a subpopulation resident within fetal tissues with immune- regulatory functions as found in MSCs. As advantage, they exhibit a greater expansion capacity and enhanced plasticity (Klemmt et al. 2011). Success of allogeneic FSC transplants in autism could be related to their great pluripotency, as well as for the absence of no expression of MHC-I and no MHC-II (Laguna Goya et al. 2011). FSCs also show strong immunomodulatory effects (Chen et al. 2011), together with stable phenotype and less senescence. Reduction in cellular ageing is a prerequisite for transplantation aims. Very important; unlike ESCs, they are not able to form teratomas posttransplantation and are obtained from tissues that would otherwise be discarded (Siniscalco et al. 2013c). Their utility in autism could be related to paracrine trophic actions on host tissues rather than cell replacement. In addition, as FSCs are derived from the three germinal layers: ectodermal, mesodermal, and endodermal layers, they retain their tissue-specific instructions and could restore dysfunctional development of the brain, gut, and immune system.

Hematopoietic stem cells (HSCs) are the cells responsible for forming blood and immune cells. They are able to modulate chronic inflammation and immune responses. As stem cells, HSCs show self-renewal, mobilization, and multipotent differentiation and homing capacities. It has been proposed that in autism these cells could be effective in restoring damaged BBB (Siniscalco et al. 2013d). Indeed, endogenous HSCs are also resident in brain and could provide constant generation of macrophagic cells without damaging BBB. Due to their role of removing cellular debris (i.e., myelin fragments), these macrophagic cells contribute to the normal homeostasis of brain function.

11.3.1 Cell Therapy in Autism: Clinical Evidences

An open label proof of concept study on the use of autologous bone marrow mononuclear cells (BMMNCs) transplantation in 32 patients with autism has recently demonstrated the efficacy of this population of cells in treating autism (Sharma et al. 2013). As the administration route was intrathecal, some adverse events (in minimal percentage) were related to the procedure (spinal headache, nausea, vomiting, backache, and pain at the site of injection); while the side effects related to cellular transplantation were transient minimal increase in hyperactivity and sometime seizures. Cell treated patients improved behaviors and sensory issue, as well as speech. Authors hypothesized that cells were able to restore hypoperfusion in brain areas related to autism.

A nonrandomized, open-label, single center phase I/II trial demonstrated the safety and efficacy of combined transplantation of human cord blood mononuclear cells (CBMNCs) and umbilical cord-derived mesenchymal stem cells (UCMSCs) in treating 37 autistic children (Lv et al. 2013). Also in this case, however, the administration route was very invasive. No adverse effects were seen. Behavior improvements were reported with the transplantation of combined cell types.

Overall these promising results are encouraging on the future use of cell therapies in treating autism. This cell therapy could be the most effective treatment for a specific autistic child, opening a new era in personalized autism management. The optimal stem cell type, dose, and administration route will need further research to achieve best results.

11.4 The Future: A Possibility for Endogenous Molecules

Recently, a new interesting study has demonstrated the efficacy of Gc protein-derived macrophage activating factor (GcMAF) in treating autistic children (Bradstreet et al. 2012). GcMAF is an endogenous protein responsible for macrophagic cell activation. In total 40 autistic children treated with GcMAF reported impressive improvements in behaviors. Several autism specialized clinicians are prescribing this molecule with very good results. By a molecular point of view, it has been demonstrated that in autism GcMAF is able to restore abnormal macrophagic cells activation via endocannabinoid system genes (Siniscalco et al. 2014). Could Gc-MAF be the future of autism care?

11.5 Conclusions

Autism is still an unexplored pathology. Several evidences highlight the plethora of biomolecular and cellular changes underlined the spectrum of disease. The research for a cure is suffering for the lack of a specific biomarker for early diagnosis of ASDs. Despite these issues, the searching for a cure is an urgent priority for both health systems and scientific world.

References

American Psychiatric Association (2000) Diagnostic and statistical manual of mental disorders (Text revision), 4th edn. American Psychiatric Association, Washington, DC

Andari E, Duhamel JR, Zalla T, Herbrecht E, Leboyer M, Sirigu A (2010) Promoting social behavior with oxytocin in high-functioning autism spectrum disorders. Proc Natl Acad Sci U S A 107(9):4389–4394

Arnold LE, Aman MG, Hollway J, Hurt E, Bates B, Li X, Farmer C, Anand R, Thompson S, Ramadan Y, Williams C (2012) Placebo-controlled pilot trial of mecamylamine for treatment of autism spectrum disorders. J Child Adolesc Psychopharmacol 22(3):198–205

Baio J (2014) Prevalence of autism spectrum disorder among children aged 8 years—autism and developmental disabilities monitoring network, 11 Sites, United States, 2010. MMWR 63(SS-2):1–22

Baribeau DA, Anagnostou E (2014) An update on medication management of behavioral disorders in autism. Curr Psychiatry Rep 16(3):437

Belsito KM, Law PA, Kirk KS, Landa RJ, Zimmerman AW (2001) Lamotrigine therapy for autistic disorder: a randomized, double-blind, placebo-controlled trial. J Autism Dev Disord 31(2):175–181

Bradstreet JJ, Vogelaar E, Thyer L (2012) Initial observations of elevated alpha-N-Acetylgalactosaminidase activity associated with autism and observed reductions from Gc protein—macrophage activating factor injections. Autism Insights 4:31–38

Brunton L, Chabner B, Knollman B (2010) Goodman and Gilman's The Pharmacological basis of therapeutics, 12th edn. McGraw Hill Professional. Berkshire, UK

Buckley AW, Sassower K, Rodriguez AJ, Jennison K, Wingert K, Buckley J, Thurm A, Sato S, Swedo S (2011) An open label trial of donepezil for enhancement of rapid eye movement sleep in young children with autism spectrum disorders. J Child Adolesc Psychopharmacol 21(4):353–357

Caplan AI, Dennis JE (2006) Mesenchymal stem cells as trophic mediators. J Cell Biochem 98(5):1076–1084

Carrasco M, Volkmar FR, Bloch MH (2012) Pharmacologic treatment of repetitive behaviors in autism spectrum disorders: evidence of publication bias. Pediatrics 129(5):e1301–1310

Chen PM, Yen ML, Liu KJ, Sytwu HK, Yen BL (2011) Immunomodulatory properties of human adult and fetal multipotent mesenchymal stem cells. J Biomed Sci 18:49

Chez MG, Aimonovitch M, Buchanan T, Mrazek S, Tremb RJ (2004) Treating autistic spectrum disorders in children: utility of the cholinesterase inhibitor rivastigmine tartrate. J Child Neurol 19(3):165–169
Chez MG, Burton Q, Dowling T, Chang M, Khanna P, Kramer C (2007) Memantine as adjunctive therapy in children diagnosed with autistic spectrum disorders: an observation of initial clinical response and maintenance tolerability. J Child Neurol 22(5):574–579
Ching H, Pringsheim T (2012) Aripiprazole for autism spectrum disorders (ASD). Cochrane Database Syst Rev 5:CD009043.
Cook EH, Leventhal BL (1996) The serotonin system in autism. Curr Opin Pediatr 8(4):348–354
Enstrom AM, Onore CE, Van de Water JA, Ashwood P (2010) Differential monocyte responses to TLR ligands in children with autism spectrum disorders. Brain Behav Immun 24(1):64–71
Essa MM, Braidy N, Vijayan KR, Subash S, Guillemin GJ (2013) Excitotoxicity in the pathogenesis of autism. Neurotox Res 23(4):393–400
Farmer CA, Aman MG (2011) Aripiprazole for the treatment of irritability associated with autism. Expert Opin Pharmacother 12(4):635–640
Farmer C, Thurm A, Grant P (2013) Pharmacotherapy for the core symptoms in autistic disorder: current status of the research. Drugs 73(4):303–314
Geier DA, Kern JK, Garver CR, Adams JB, Audhya T, Geier MR (2009) A prospective study of transsulfuration biomarkers in autistic disorders. Neurochem Res 34(2):386–393
Gentile S (2014) Risks of neurobehavioral teratogenicity associated with prenatal exposure to valproate monotherapy: a systematic review with regulatory repercussions. CNS Spectr 26:1–11 (Epub ahead of print)
Ghosh A, Michalon A, Lindemann L, Fontoura P, Santarelli L (2013) Drug discovery for autism spectrum disorder: challenges and opportunities. Nat Rev Drug Discov 12(10):777–790
Greenaway M, Elbe D (2009) Focus on aripiprazole: a review of its use in child and adolescent psychiatry. J Can Acad Child Adolesc Psychiatry 18(3):250–260
Guastella AJ, Einfeld SL, Gray KM, Rinehart NJ, Tonge BJ, Lambert TJ, Hickie IB (2010) Intranasal oxytocin improves emotion recognition for youth with autism spectrum disorders. Biol Psychiatry 67(7):692–694
Hardan AY, Handen BL A retrospective open trial of adjunctive donepezil in children and adolescents with autistic disorder (2002) J Child Adolesc Psychopharmacol 12(3):237–241
Hollander E, Bartz J, Chaplin W, Phillips A, Sumner J, Soorya L, Anagnostou E, Wasserman S (2007) Oxytocin increases retention of social cognition in autism. Biol Psychiatry 61(4):498–503
Hollander E, Chaplin W, Soorya L, Wasserman S, Novotny S, Rusoff J, Feirsen N, Pepa L, Anagnostou E (2010) Divalproex sodium vs placebo for the treatment of irritability in children and adolescents with autism spectrum disorders. Neuropsychopharmacology 35(4):990–998
Hviid A, Melbye M, Pasternak B (2013) Use of selective serotonin reuptake inhibitors during pregnancy and risk of autism. N Engl J Med 369(25):2406–2415
http://www.fda.gov/Safety/MedWatch/SafetyInformation/SafetyAlertsforHumanMedicalProducts/ucm152595.htm
Insel TR (2010) The challenge of translation in social neuroscience: a review of oxytocin, vasopressin, and affiliative behavior. Neuron 65(6):768–779
Jacob S, Brune CW, Carter CS, Leventhal BL, Lord C, Cook EH Jr (2007) Association of the oxytocin receptor gene (OXTR) in Caucasian children and adolescents with autism. Neurosci Lett 417(1):6–9
Karst JS, Van Hecke AV (2012) Parent and family impact of autism spectrum disorders: a review and proposed model for intervention evaluation. Clin Child Fam Psychol Rev 15(3):247–77
Kern JK, Geier DA, Adams JB, Garver CR, Audhya T, Geier MR (2011) A clinical trial of glutathione supplementation in autism spectrum disorders. Med Sci Monit 17(12):CR677–682
Kikuchi T, Tottori K, Uwahodo Y, Hirose T, Miwa T, Oshiro Y, Morita S (1995) 7-(4-[4-(2,3-Dichlorophenyl)-1-piperazinyl]butyloxy)-3,4-dihydro-2(1H)-quinolinone (OPC-14597), a new putative antipsychotic drug with both presynaptic dopamine autoreceptor

agonistic activity and postsynaptic D2 receptor antagonistic activity. J Pharmacol Exp Ther 274(1):329–336
King BH, Hollander E, Sikich L, McCracken JT, Scahill L, Bregman JD, Donnelly CL, Anagnostou E, Dukes K, Sullivan L, Hirtz D, Wagner A, Ritz L (2009) STAART psychopharmacology network. Lack of efficacy of citalopram in children with autism spectrum disorders and high levels of repetitive behavior: citalopram ineffective in children with autism. Arch Gen Psychiatry 66(6):583–590
Klemmt PA, Vafaizadeh V, Groner B (2011) The potential of amniotic fluid stem cells for cellular therapy and tissue engineering. Expert Opin Biol Ther 11(10):1297–1314
Laguna Goya R, Busch R, Mathur R, Coles AJ, Barker RA (2011) Human fetal neural precursor cells can up-regulate MHC class I and class II expression and elicit CD4 and CD8T cell proliferation. Neurobiol Dis 41(2):407–414
Li L, Jiang J (2011) Regulatory factors of mesenchymal stem cell migration into injured tissues and their signal transduction mechanisms. Front Med 5(1):33–39
Lv YT, Zhang Y, Liu M, Qiuwaxi JN, Ashwood P, Cho SC, Huan Y, Ge RC, Chen XW, Wang ZJ, Kim BJ, Hu X (2013) Transplantation of human cord blood mononuclear cells and umbilical cord-derived mesenchymal stem cells in autism. J Transl Med 11:196
Martin HG, Manzoni OJ (2014) Late onset deficits in synaptic plasticity in the valproic acid rat model of autism. Front Cell Neurosci 8:23
McGinnis WR (2004) Oxidative stress in autism. Altern Ther Health Med 10(6):22–36
McKinney C, Renk K (2011) Atypical antipsychotic medications in the management of disruptive behaviors in children: safety guidelines and recommendations. Clin Psychol Rev 31(3):465–471
McPherson RA, Hardy G (2011) Clinical and nutritional benefits of cysteine-enriched protein supplements. Curr Opin Clin Nutr Metab Care 14(6):562–568
Mostafa GA, El-Hadidi ES, Hewedi DH, Abdou MM (2010) Oxidative stress in Egyptian children with autism: relation to autoimmunity. J Neuroimmunol 219(1–2):114–118
Nicolson R, Craven-Thuss B, Smith J (2006) A prospective, open-label trial of galantamine in autistic disorder. J Child Adolesc Psychopharmacol 16(5):621–629
Niederhofer H (2007) Treating autism pharmacologically: also tacrine might improve symptomatology in some cases. J Child Neurol 22(8):1054
Owley T, Salt J, Guter S, Grieve A, Walton L, Ayuyao N, Leventhal BL, Cook EH Jr (2006) A prospective, open-label trial of memantine in the treatment of cognitive, behavioral, and memory dysfunction in pervasive developmental disorders. J Child Adolesc Psychopharmacol 16(5):517–524
Patterson PH (2011) Modeling autistic features in animals. Pediatr Res 69(5 Pt 2):34R–40R
Rai D, Lee BK, Dalman C, Golding J, Lewis G, Magnusson C (2013) Parental depression, maternal antidepressant use during pregnancy, and risk of autism spectrum disorders: population based case-control study. BMJ 346:f2059
Remington G, Sloman L, Konstantareas M, Parker K, Gow R (2001) Clomipramine versus haloperidol in the treatment of autistic disorder: a double-blind, placebo-controlled, crossover study. J Clin Psychopharmacol 21(4):440–444
Scott LJ, Dhillon S (2008) Spotlight on risperidone in irritability associated with autistic disorder in children and adolescents. CNS Drugs 22(3):259–262
Seeman P, Caruso C, Lasaga M (2008) Memantine agonist action at dopamine D2High receptors. Synapse 62(2):149–153
Sharma A, Gokulchandran N, Sane H, Nagrajan A, Paranjape A, Kulkarni P, Shetty A, Mishra P, Kali M, Biju H, Badhe P (2013) Autologous bone marrow mononuclear cell therapy for autism: an open label proof of concept study. Stem Cells Int 2013:623875
Siniscalco D (2012) Stem cell research: an opportunity for autism spectrum disorders treatment. Autism 2:e106
Siniscalco D (2013) Current findings and research prospective in autism spectrum disorders. Autism S2:e001

Siniscalco D, Antonucci N (2013a) Possible use of Trichuris suis ova in autism spectrum disorders therapy. Med Hypotheses 81(1):1–4
Siniscalco D, Antonucci N (2013b) Involvement of dietary bioactive proteins and peptides in autism spectrum disorders. Curr Protein Pept Sci 14(8):674–679
Siniscalco D, Cubala-Kucharska M (2013) Treatment of the child with autism-newest medical trends. Autism 3:110
Siniscalco D, Giordano C, Galderisi U, Luongo L, Alessio N, Di Bernardo G, de Novellis V, Rossi F, Maione S (2010) Intra-brain microinjection of human mesenchymal stem cells decreases allodynia in neuropathic mice. Cell Mol Life Sci 67(4):655–669
Siniscalco D, Giordano C, Galderisi U, Luongo L, de Novellis V, Rossi F, Maione S (2011) Long-lasting effects of human mesenchymal stem cell systemic administration on pain-like behaviors, cellular, and biomolecular modifications in neuropathic mice. Front Integr Neurosci 5:79
Siniscalco D, Giordano A, Galderisi U (2012a) Novel insights in basic and applied stem cell therapy. J Cell Physiol 227(5):2283–2286
Siniscalco D, Sapone A, Cirillo A, Giordano C, Maione S, Antonucci N (2012b) Autism spectrum disorders: is mesenchymal stem cell personalized therapy the future? J Biomed Biotechnol 2012:480289
Siniscalco D, Sapone A, Giordano C, Cirillo A, de Novellis V, de Magistris L, Rossi F, Fasano A, Maione S, Antonucci N (2012c)The expression of caspases is enhanced in peripheral blood mononuclear cells of autism spectrum disorder patients. J Autism Dev Disord 42(7):1403–1410
Siniscalco D, Cirillo A, Bradstreet JJ, Antonucci N (2013a) Epigenetic findings in autism: new perspectives for therapy. Int J Environ Res Public Health 10(9):4261–4273
Siniscalco D, Di Marsilio A, Antonucci N (2013b) Ethics in autism care. Autism 3:e119
Siniscalco D, Bradstreet JJ, Sych N, Antonucci N (2013c) Perspectives on the use of stem cells for autism treatment. Stem Cells Int 2013:262438.
Siniscalco D, Bradstreet JJ, Antonucci N (2013d) Therapeutic role of hematopoietic stem cells in autism spectrum disorder-related inflammation. Front Immunol 4:140.
Siniscalco D, Bradstreet JJ, Cirillo A, Antonucci N (2014) The in vitro GcMAF effects on endocannabinoid system transcriptionomics, receptor formation, and cell activity of autism-derived macrophages. J Neuroinflammation 17(11):78.
Stigler KA, Mullett JE, Erickson CA, Posey DJ, McDougle CJ (2012) Paliperidone for irritability in adolescents and young adults with autistic disorder. Psychopharmacology (Berl) 223(2):237–245
Troost PW, Lahuis BE, Steenhuis MP, Ketelaars CE, Buitelaar JK, van Engeland H, Scahill L, Minderaa RB, Hoekstra PJ (2005) Long-term effects of risperidone in children with autism spectrum disorders: a placebo discontinuation study. J Am Acad Child Adolesc Psychiatry 44(11):1137–1144
University of Maryland Medical Center Autism. http://umm.edu/health/medical/ency/articles/autism#ixzz2wPdFF8MI. Accessed 1 April 2015
Wasserman S, Iyengar R, Chaplin WF, Watner D, Waldoks SE, Anagnostou E, Soorya L, Hollander E (2006) Levetiracetam versus placebo in childhood and adolescent autism: a double-blind placebo-controlled study. Int Clin Psychopharmacol 21(6):363–367
Wermter AK, Kamp-Becker I, Hesse P, Schulte-Körne G, Strauch K, Remschmidt H (2010) Evidence for the involvement of genetic variation in the oxytocin receptor gene (OXTR) in the etiology of autistic disorders on high-functioning level. Am J Med Genet B Neuropsychiatr Genet 153B(2):629–39
Williams K, Brignell A, Randall M, Silove N, Hazell P (2013) Selective serotonin reuptake inhibitors (SSRIs) for autism spectrum disorders (ASD). Cochrane Database Syst Rev 8:CD004677
Witschi A, Reddy S, Stofer B, Lauterburg BH (1992) The systemic availability of oral glutathione. Eur J Clin Pharmacol 43(6):667–669
Zablotsky B, Kalb LG, Freedman B, Vasa R, Stuart EA (2014) Health care experiences and perceived financial impact among families of children with an autism spectrum disorder. Psychiatr Serv 65(3):395–398

Index

M. d. l. A. Robinson-Agramonte (ed.), *Translational Approaches to Autism Spectrum Disorder*, DOI 10.1007/978-3-319-16321-5

MIX
Papier aus verantwortungsvollen Quellen
Paper from responsible sources
FSC® C105338
FSC
www.fsc.org

If you have any concerns about our products,
you can contact us on
ProductSafety@springernature.com

In case Publisher is established outside the EU,
the EU authorized representative is:
Springer Nature Customer Service Center GmbH
Europaplatz 3, 69115 Heidelberg, Germany

Printed by Libri Plureos GmbH
in Hamburg, Germany